Towards Conservation of Omani Local Chicken:
Management, Performance and Genetic Diversity

Towards Conservation of Omani Local Chicken: Management, Performance and Genetic Diversity

Dissertation
to obtain the Ph.D. degree
in the International Ph.D. program for Agricultural Sciences in Göttingen (IPAG)
at the Faculty of Agricultural Sciences,
Georg-August-Universität Göttingen, Germany

presented by
Badar Al-Qamashoui
Born in Ibri, Sultanate of Oman

Department of Animal Sciences
Animal Husbandry in the Tropics and Subtropics

Göttingen, June 2014

Bibliografische Information der Deutschen Nationalbibliothek
Die Deutsche Nationalbibliothek verzeichnet diese Publikation in der Deutschen Nationalbibliografie; detaillierte bibliografische Daten sind im Internet über http://dnb.d-nb.de abrufbar.
1. Aufl. - Göttingen : Cuvillier, 2014
Zugl.: Göttingen, Univ., Diss., 2014

Nonnenstieg 8, 37075 Göttingen
Telefon: 0551-54724-0
Telefax: 0551-54724-21
www.cuvillier.de

1. Auflage, 2014
Gedruckt auf umweltfreundlichem, säurefreiem Papier aus nachhaltiger Forstwirtschaft.

ISBN 978-3-95404-733-8
eISBN 978-3-7369-4733-7

D 7

1. Supervisor: Prof. Dr. Eva Schlecht
   Animal Husbandry in the Tropics and Subtropics
   Universität Kassel and Georg-August Universität Göttingen

2. Co-Supervisor: Prof. Dr. Henner Simianer
   Animal Breeding and Genetics
   Georg-August Universität Göttingen

In collaboration with:
Dr. Steffen Weigend
Institute of Farm Animal Genetics
Friedrich- Loeffler-Institut (FLI)
Neustadt-Mariensee, Germany

Dr. Aliya Al-Ansari
Department of Biology
Sultan Qaboos University (SQU)
Muscat, Oman

Date of the oral exam: February 10, 2014

Table of contents

Acknowledgments

First of all, I would like to express my sincere gratitude to my advisor Prof. Dr. Eva Schlecht for the continuous support of my PhD study and research, for her patience, understanding and enthusiasm. Her guidance and motivation helped me during these past five years. My grateful thanks to Prof. Dr. Henner Simianer for taking over co-referee and for accepting this thesis. Besides my advisors, I would like to thank Dr. Steffen Weigend for welcoming me to his laboratory and for valuable discussions, enlighting comments on the manuscripts and hard questioning during my presentations.

I thank Prof. Dr. Osman Mahgoub Gaafar from Sultan Qaboos University for his valuable assistance during my field visits, sampling and lab work in Oman. My sincere thanks also go to Dr. Alia Al-Ansari for supervising me in the third part of my project. I am also grateful to Dr. Albano Beja Pereira from University of Porto for his fruitful ideas and discussions and for teaching me many software applications and giving useful suggestions concerning my topic.

I highly appreciate the technical assistance of Mrs. Annett Weigend, Mr. Maik Przyklenk and Mrs. Natalie Janus at the Institute of Farm Animal Genetics, Mariensee. I, also, thank my fellow lab-mate in Germany, Mr. Charles Limo, for inspiring discussions during the sleepless nights when we were working together in Mariensee, and for all the fun we have had in the past four years. I would like to extend my thanks to Mr. Farhad Vahidi for his help in learning and running some software programmes in population genetics during his stay in Göttingen.

Special thanks go to everyone who provided valuable assistance and cooperation during sampling and also to farmers who allowed me to collect blood samples from their chicken and for sharing their knowledge regarding their farm management. My thanks extend to Mr. Hilal Al-Shikaili and Mrs. Fatma Al-Salmi for lab assistance at Sultan Qaboos University.

I would also like to thank the secretarial staff for helping me any time I needed it and for offering me a warm reception - Mrs. Birsen Iffländer and Mrs. Uta Döring. I will forever be thankful to Mrs. Grete Thinggaard for answering all my questions concerning my study in Göttingen.

I am greatly thankful to the Ministry of Agriculture and Fisheries for providing me a full time leave so as to complete my study in Germany, and also to the Ministry of Higher Education for supporting me with a part-time scholarship.

Last but not least, I would like to thank my family: my parents, for their continuous prayers in supporting me spiritually, throughout my life, my lovely wife and my four kids for their patience and sympathy during the time I was away from home. My wife was always giving me advice and encouragement during my study period. I would not have made it this far without her support.

List of figures

List of tables

Summary

Many rural families in Oman are engaged in agricultural and animal husbandry activities with a majority still depending on farming as a main source of income. Local chicken farming in Oman represents one of the main agricultural activities that provide opportunities for food security and income for many rural families. Despite its importance, there is no detailed study for evaluating the production system, production performance and genetic potential of local poultry in Oman. The present thesis aimed at:

1. Characterizing the local chicken management, production and marketing strategies of small-scale farming;
2. Assessing the production traits and phenotypic features of Omani local chickens;
3. Evaluating the genetic makeup and diversity, and assess the conservation possibilities for traditional chicken types in Oman;
4. Contributing to tracing the maternal origins of chicken populations in Oman as well as in the Arabian Peninsula.

The present thesis consists of three studies. In the first study, a structured questionnaire was used to collect data from 163 households distributed across 18 villages in Oman's six major agro-ecological zones. These were: Batinah (BT), Dhofar (DF), North Hajar (NH), East Hajar (EH), Musandam (MU), and East Coast (EC). Free-range scavenging was the dominant production system, but 58.5% of the respondents offered commercial feed supplements to their chicken. The purposes of chicken keeping were: egg production for domestic use (69%) and income generation (31%). Omani local chickens widely vary in plumage color patterns, comb types, shank colors and other phenotypic characteristics. Male and female body weight also varied, being 1.34 ±0.65 kg and 1.14 ±0.86 kg ($P<0.05$), respectively. Flock size averaged 22 ±7.7 chickens per household with 4.8 hens per one cock. Clutch size was 12.3 ±2.85 eggs and annual egg production averaged 64.5 ±2.85 eggs per hen. Egg hatchability was 88% ±6.0, and annual chicken mortality across all age and sex categories was 16% ±1.4. Predators were the major production constraint (26.5%), followed by high feed prices, low egg production and low chicken body growth. Logistic and multiple regression analysis showed that several socio-economic factors of chicken owners influenced feeding, housing, and health care of the chicken ($P<0.05$). The strong involvement of women makes them key stakeholders in future development and conservation programs of local chicken.

In the second study, twenty-nine microsatellite markers were used on 158 birds from the above six agro-ecological zones. Across loci and populations, a total of 217 alleles were observed. Across populations, the average number of alleles per locus was 7.48 and ranged from 2 (MCW98 and MCW103) to 20 (LEI094). Across populations, the mean expected heterozygosity (H_E) was 0.62. The mean global deficit of heterozygotes across populations (F_{IT}) was 0.159 while average fixation index (F_{ST}) between populations was 0.034, indicating a low population differentiation. Based on Nei's genetic distance a neighbor-joining tree was constructed for the populations, which clearly identified the Dhofar population as the most distant one of the Omani chicken populations. The analysis of conservation priorities identified DF and MU populations as the ones that largely contribute to the maximal genetic diversity of the Omani chicken gene pool.

In the third study, sequencing data from a fragment of the control region of mitochondrial genome (mtDNA) from 175 individuals and 32 published sequences was used to assess genetic diversity and inference on the maternal origins of local chickens from the Arabian Peninsula (Oman, Saudi Arabia, Yemen, Isle of Socotra) and the Horn of Africa. Because of its role in the human movements between Asia and Africa and to investigate the dispersal of chicken around the Indian Ocean Rim, sequences from Africa and India were also included in this study. We found a total of 27 haplotypes with an average haplotype diversity of 0.7588 ±0.0300, clustering into three of the previously identified phylogenetic clades. The most frequent observed haplotypes from the Arabian Peninsula (and Socotra) clustered in clade E, which is supposed to have originated on the Indian subcontinent. While samples from Somalia belong mostly to clade C, which supposedly has its roots in Southeast Asia, a few individuals, mostly from North Oman, clustered in clade A, originating from Southeast and/or East Asia. The wide presence of clade E on the Arabian Peninsula points towards a major influence of the Indus Valley as center of origin in the genesis of Arabian local chicken. Isolation by distance tests showed that chicken diffusion across the Indian Ocean is correlated with the proximity to the main centers of chicken domestication. The high frequency of haplotypes originating from the Indian Subcontinent domestication event, on the Arabian Peninsula, provides interesting insight into the role of the Peninsula in the diffusion of livestock around the Indian rim.

CHAPTER 1

General introduction

1.1 Background

With an area of 309,500 km^2, Oman is the third largest country on the Arabian Peninsula. It is located in the southeastern part between latitudes 16°40' and 26°20' north and longitudes 51°50' and 59°40' east with a coastline extending for 3,165 km (DGMAN 2012). The Sultanate borders the Kingdom of Saudi Arabia in the West, the United Arab Emirates in the Northwest, the Republic of Yemen in the South, the Strait of Hormuz in the North and the Arabian Sea in the East.

Oman is generally an arid subtropical country with two distinct seasons: winter from November to April, and summer from May to October (Al-Mashakhi and El-Hag 2007). With the exception of some higher altitudes in the Interior and remote South, the climate in summer is hot and dry with a full-day average temperature of 38°C, whereas in the winter temperatures are mild (15-23°C). The precipitation is generally low and irregular, especially in the Interior region, with an average of 117.4 mm/year for the whole country (DGMAN 2012).

The country has a varied topography, including mountain ranges, arid deserts and fertile plains. The wide variation in climatic and landscape features is the reason for the country's abundant and unique faunal and floral biodiversity in the different agro-ecological zones (AEZ) (Al-Zidjali 1996; Al-Saadi 2013). Oman is separated into several agro-ecological zones (Table 2.1) based on topography and climate, parameters which influence crop water requirements and efficient use of water, land and water resources and cropping patterns (Al-Zidjali 1996).

Agriculture is an important economic sector and plays a crucial role for the food security objective of the Sultanate of Oman. The size of the cultivated area is 73,670 hectares (DGALR 2011) and around 40% of the population is still engaged in the agricultural sector (MoNE 1995). According to the target set for the agriculture sector in the *'Vision 2020'*, its contribution to GDP is expected to rise to 3.1% by 2020 with an annual growth of not less than 4.5% (CBO 2011). Date palm, banana, mango, coconut, vegetables and fodder and field crops are the major agricultural products and considered as the main plant genetic resources. In addition, the country possess indigenous grasses, medicinal plants, pastures, trees and shrubs, and forest resources (DGALR 2011).

Livestock production is a central farming activity in Oman. The total number of livestock in the country is around 2.5 million, composed of goats (1,685,420), sheep (380,990), cattle (326,240) and camels (127,010) (DGALR 2011). The majority of cattle and camels are in the most southern region of Dhofar whereas the majority of goats and sheep are kept in the Batinah plain. Non-official information from the 2013 Agricultural Census estimated the total number of local chickens in Oman as 2.4 million (personal communication; Dr. Khalid Alzadjali, Ministry of Agriculture and Fisheries).

Livestock farming has been practiced in Oman for thousands of years although its history remains debated. Ancient cave drawings (Figure S1.1) (AbdulNayeem 2000) and excavated bone remains from animals in Oman (Bokonyi 1992) and the Arabian Peninsula (Groucutt and

Petraglia 2012) support the presence of domesticated animals in the Late Stone Age period (Bokonyi 1992; Wilkens 2005). Other scholars, however, defend that maritime-oriented fishing cultures have appeared along the coast of Oman as early as the 7th millennium BP (Biagi 1994), giving rise to preliminary farming settlements. Discoveries of a first domestication event of dromedary camel in the south of Oman (Grigson et al. 1989; Zeder et al.2006) and of horses in Saudi Arabia (SCTA 2013) around 5000 BP, have given further evidence for very early domestication activities. Being at a very important and stratigic geographical location, Boivin et al. (2010) suggested that the ancient Arabian Peninsula played an important role in channeling plants, crops and animals between their centers of origin and their areas of dispersion. Besides the strong oceanic trade routes via the Arabian Sea and the Indian Ocean (Fuller et al. 2011), evidences for earliest inland trade routes from South to North of the Peninsula using camels have also been documented (Pickering 2007).

Many conservation and improvement programs for local livestock breeds, ranging from short- to long-term experiments, have been conducted by the Ministry of Agriculture and Fisheries of Oman (DGALR 2011). The main species targeted are cattle, sheep, goat and chickens. However, these programs lack studies analyzing the molecular genetic makeup of these species. Recently, the Ministry, in the framework of a national conservation strategy, has decided to update its programs by adopting the procedures recommended by the Food and Agriculture Organization of the United Nations, i.e. by using advanced genetic tools in conservation approaches (DGALR 2011).

1.2 Genetic diversity in livestock and role of conservation

Genetic diversity is defined as the variety of alleles and genotypes present in a population that is reflected in morphological, physiological and behavioral differences between individuals and populations (Frankham et al. 2002; Delany 2003). Local farm animals are an important reservoir of genetic diversity as it is essential to meet their current production needs in various environments and to facilitate rapid adaptation to changing breeding objectives (Notter 1999). However, the loss of genetic diversity within these farm animals has become a major concern in the last decades. Many indigenous breeds that have unique characteristics such as disease resistance and adaptation to their environment are being replaced by industrial breeds (Perera 2010; FAO 2012). Around 22% of the world's livestock breeds are classified as being at risk of extinction, due to loss of genetic diversity and decrease in population sizes by crossbreeding with commercial exotic breeds (FAO 2012). These specialized exotic breeds in many livestock species now suffer from the consequences of inbreeding, and as a result, many productive breeds are becoming more dependent on intensive management (Wollny 2003; Gibson et al. 2005). There is a need, therefore, to slow down the degradation of farm animal genetic resources and establish programs for their conservation and sustainable use (Gibson et al. 2005; Perera 2010).

Conserving programs aim to preserve valuable genetic resource in order to face any future environmental changes or disasters (Allendorf and Luikart 2007). They also aim to reserve these populations as a source of rare alleles and contribute to the search for genes associated with health and quality traits (Gandini and Oldenbroek 1999; Mendelsohn 2003). In many cases conservation programs are structured to avoid inbreeding and conserve the observed phenotypic differences and genetic variation within the different lines (Marle-Koster and Nel 2003).

Several conservation options and strategies have been established (Gibson et al. 2005). Among these, the strategy which takes into account both within- and between- subpopulation components of coancestry is recommended (Caballero and Toro 2002; Ollivier and Foulley 2005; Fernandez et al. 2008). This approach has been used to determine the optimal contribution of each subpopulation in a synthetic population or gene pool of maximum gene or allelic diversity (Perez-Figueroa et al. 2009). Estimating these optimal contributions can be applied to prioritize subpopulations for conservation (Caballero and Toro 2002; Perez-Figueroa et al. 2009). Caballero and Toro (2002) stated that the procedure of contributions of minimum coancestry has been shown to maximize the genetic diversity of the population in terms of expected heterozygosity and effective population size. This approach has also been shown to be very effective in preserving the original distribution of allelic frequencies in conservation programs (Saura et al. 2008) and maintains to a certain extent the allelic richness of the population (Fernandez et al. 2008). Efficient conservation programs require a good knowledge of the genetic structure of these local populations, as well as an assessment of their diversity at the molecular level to provide recommendations regarding their future management (Boettcher et al. 2010).

1.3 Local chicken breeds: production system and genetic diversity assessment

Local "indigenous" chickens play a crucial role for the livelihood of most rural families in the developing world. Besides providing food, local chickens are important for income generation. Most rural families in developing countries are involved in local chicken husbandry due to its low capital investments (Jens et al. 2004; Gibson et al. 2005). Local chicken husbandry is frequently under the responsibility of women involved in most poultry management operations (Mwalusanya et al. 2002). The majority of households in these communities lacks the required husbandry skills, training and market opportunities to effectively improve animal production (Barua and Yoshimur 1997; Mwalusanya et al. 2002; IAEA 2004; Pica-Ciamarra and Dhawan 2010).

Free-range scavenging system is the main production system in the tropics and subtropics (Aini 1990; Barua and Yoshimur 1997; Dessie and Ogle 2001). Under this production system, local chicken flocks are managed extensively, which enables them to obtain most of their feed through scavenging. Local chicken types are characterized by considerable phenotypic variation (Mcainsh et al. 2004). They are considered as an important genetic reservoir that developed under harmful environmental conditions, diseases and predators. Their long adaptation to this harsh environments enables them to resist extreme temperatures, poor nutrition and absence of veterinary care, and in turn survive and reproduce (Hall 1986). It is assumed that by raising chickens under these harsh environmental conditions, diverse allele and allele combinations will be produced through natural selection that gave these breeds adaptation and a reasonable ability to produce (Horst et al. 1996).

Local poultry breeds in many countries have provided an interesting alternative to commercial strains, providing typical products with particular meat qualities that are of great interest to the regional local markets (Zanetti 2009). Commercial chicken purebreds were selected for performance traits and managed as closed populations with well documented pedigrees and breeding history. Commercial poultry breeds have been selected to be reared in an optimum feeding system and therefore, scavenging conditions may not satisfy their nutritional needs (Leroy et al. 2012). Consequently, efforts for conserving local chicken are of greatest importance as they allow breeders to take advantage of unique adaptive traits present in this diversity that enables them to respond to changes in the environment (Besbes et al. 2007).

Molecular genetic markers have been widely used as tools to study the genetic diversity and to design conservation and breeding programs for local populations. A marker is an identified genome site that exhibits polymorphism (Beuzen et al. 2000). Among different molecular markers, microsatellites have been extensively used to describe the genetic diversity in many livestock species. Microsatellites are short DNA stretches consisting of a repeat motif of usually a two- or four-nucleotide sequence, also known as simple tandem repeats. They are characterized by their wide distribution in the genome, easy to use and highly polymorphic (Cheng and Muir 2005). Microsatellites can be amplified for identification by the polymerase chain reaction (PCR) process, using the unique sequences of flanking regions as primers (Beuzen et al. 2000; Cheng and Muir 2005). This process results in the production of enough DNA to be visible on agarose or polyacrylamide gels.

Many genetic diversity variables and approaches can be achieved by using microsatellites. These include allele frequencies, private alleles, proportions of polymorphic loci, observed and expected heterozygosity, phylogenetic relationships, genetic admixtures and population structures (Chikhi and Bruford 2005). Genetic differentiation among populations has been assessed using genetic distance measures such as Nei's (Nei and Li 1979), and Reynolds (Reynolds et al. 1983) genetic distances. Microsatellites have also been used in identifying genetically important populations for conservation (Bennewitz and Meuwissen 2005).

In chickens, microsatellite DNA typing has been extensively used for genotyping chicken (Romanov and Weigend 2001; Hillel et al. 2003; Granevitze et al. 2007). A set of 30 microsatellite markers in chickens has been recommended by the Food and Agriculture Organization, including ADL0268, ADL0278, ADL0112, LEI0192, LEI0234, LEI0094, LEI0166, MCW0206, MCW0295, MCW0081, MCW0014, MCW0183, MCW0067, MCW0104, MCW0123, MCW0330, MCW0165, MCW0069, MCW0248, MCW0111, MCW0020, MCW0034, MCW0103, MCW0222, MCW0016, MCW0037, MCW0098, MCW0284, MCW0078 and MCW0216 (FAO 2004).

Appling microsatellite genotyping on indigenous chicken breeds, showed high genetic diversity levels of these breeds (Muchadeyi et al. 2007; Mwacharo et al. 2007; Shahbazi et al. 2007; Berthouly et al. 2008; Cuc et al. 2010; Mtileni et al. 2011b). In most cases, no clear substructuring has been observed among local chicken ecotypes across distant agro-ecological zones (Muchadeyi et al. 2007; Mtileni et al. 2011b), while their clear isolation from commercial breeds was detected in many studies (Muchadeyi et al. 2007; Leroy et al. 2012).

Another important type of markers is the DNA of mitochondria (mtDNA), which is an extra-nuclear genetic material that has been widely used for analyses of genetic diversity. The avian mtDNA is a double-stranded circular molecule that is 16,775 bp in size (Desjardins and Morais 1990). The highly polymorphic displacement loop region (control region) of the mtDNA (1 – 1232 bp) contains the elements that control the replication of the molecule (Akishinonomiya et al. 1994). The control region of mtDNA has been used by many researchers in the past decade, particularly as a means of locating individual domestication centers and the routes of subsequent dispersals (Miao et al. 2013).

The mtDNA has a maternal mode of inheritance with absence of recombination, therefore, it became an ideal marker for phylogenetic studies and to trace the geographic distribution of species (Galtier et al. 2009). It is assumed that the existence of multiple mtDNA lineages and their mixing within breeds could be due to multiple domestication events or to introgression between domestic and wild species (Galtier et al. 2009). Therefore it can be inferred that animals that share similar mtDNA must have a common female ancestor.

Many studies based on mtDNA have contributed to the current understanding of the geographic distribution and origin of domestic chicken across various regions of the world (e.g., Liu et al. 2006; Oka et al. 2007; Gongora et al. 2008; Razafindraibe et al. 2008). Chicken have been deeply integrated into the human culture as early as 5400 BC (West and Zhou 1988). It was first suggested that domestic chicken (*Gallus gallus domesticus*) has been domesticated from red jungle fowl (*Gallus gallus*) in southeast Asia or in the Yellow River valley (Fumihito et al. 1994). However, a recent study has proved that besides *Gallus gallus* several other species and subspecies from Yunnan, South and Southwest China and/or surrounding areas (i.e., Vietnam, Burma, and Thailand), and the Indian subcontinent, also contributed to the genesis of modern chicken (Nishibori et al. 2005; Liu et al. 2006; Miao et al. 2013). Liu et al. (2006) in particular, studied the mtDNA D-loop segment in a large and diverse gene pool of domestic chickens from a wide geographic area (Europe and Asia), and suggested, for the first time, nine clades (A-I) representing the main maternal lineages of modern domesticated chickens.

As the genetic maternal lineages of chicken are very well characterized with respect to their geographical origins, the use of mtDNA to assess the origins of chickens in any part of the world seems very promising (Liu et al. 2006; Galtier et al. 2009). In eastern Africa for instance, a likely Indian subcontinent origin for the commonest haplogroup of domestic village chickens have been reported (Muchadeyi et al. 2008; Mtileni et al. 2011a; Mwacharo et al. 2011). It was suggested that the coastal maritime trading networks around the Indian Ocean were the main routes for the introduction of chicken in Eastern Africa (Mwacharo et al. 2011). Among several scenarios, the scenario of an Arabian Peninsula involvement in the introduction of chicken, as in cattle (Hanotte et al. 2002), into the Horn of Africa and East Africa has been suggested (Mwacharo et al. 2013_a; Mwacharo et al. 2013_b).

Using both microsatellites and mtDNA markers could be a complementary approach in assessing the genetic diversity of local chickens. Evaluating the polymorphism patterns of both sets of markers with different modes of inheritance will allow tracking more recent demographic events along with phylogeographic events dating further back in time (Cuc 2010). Therefore, combining both markers can provide more insights into the evolutionary forces determining the genetic makeup of livestock breeds.

1.4 Risks of extinction of Omani local chickens

Local chicken production is one of the farming activities in the rural communities of Oman (MAF 2013). Only few reports have been published about the local chicken production system in Oman. Omani local chicken are characterized by their small size body and large variation in plumage color (Kadim et al. 2009). The name *mahalli* (local) was given to the local chickens, considering them as one population.

Despite the existence of the commercial industry, the local chicken lines are found in most villages especially in remote rural areas, where they contribute partially to household food consumption and production (Saleh 2000; Kadim et al. 2009; MAF 2013). Importation of exotic chicken breeds for commercial investments has gradually increased during the past years due to the high local demand on chicken products. There were 24,730,000 commercial layers and broilers spread over the country. The production of poultry meat produced on commercial farms has doubled within two years (2008-2010), and the production of eggs reached 183 million in 2010 (DGALR 2011).

The above figures indicate a major role of exotic chickens in the country and point to the danger of continuous gene flow and genetic erosion of local chicken genetic resources. The replacement of local by exotic breeds and/or uncontrolled mixing with local populations has been posing a serious threat to the existence of local chicken breeds on small-scale farms, putting these local animal genetic resources at risk of extinction (Saleh 2000; Kadim et al. 2009).

Several extension programs have been conducted in an attempt to improve the chicken production in Oman. Among these, a more recent extension program targeting the local chicken sector in Oman is the Small-scale Local Chicken Units (SLCU) that has been introduced by the Directorate of Rural Women Development of the Ministry of Agriculture and Fisheries (MAF 2013). The program aims to improve the production performance of smallholder local chicken through applying advanced housing and feeding conditions and management assets. Other goals of this program were to improve the income and nutritional status of rural families and to contribute to rural development through more holistic and self-reliant approaches. In its first stage, 326 small-scale chicken units (50 birds capacity; Figure S1.2) have been constructed for the beneficiaries with provision of extension services such as feeding and laying assets. In parallel, a set of short visits and workshops for the targeted chicken owners are offered.

For genetic conservation and improvement purposes, two long-term genetic research programs for local chickens have been conducted in Oman; one at the Ministry of Agriculture and Fisheries Research Center in Dhofar (started in 1992) and one at the Animal Research Station of Sultan Qaboos University in Muscat (started in 2002). In both projects, local chicken flocks were randomly selected from villages and kept at the research units, where they were subjected to selection programs based on the number and weight of eggs produced during a period of 52 weeks; the program also included the birds' performance in the selection index of the second generation (Saleh 2000; Kadim et al. 2009).

The ongoing extensional and conservation programs lack information about the genetic makeup, diversity and structure of chickens in Oman. It is not clear whether local chicken in different agro-ecological zones of Oman form distinct genetic populations. Assessment of the genetic makeup of chicken populations can help in determining their priorities for conservation. The sales prices of local chicken and their products are higher than that of products from commercial lines (Kadim et al. 2009). With their high consumer preferences (Saleh 2000; Kadim et al. 2009), and being an elementary part of Omanis' diets, local chickens are considered as strategic sources for food security and as a valuable asset for the country's genetic resources (DGALR 2011).

1.5 Scope of the thesis

Taking into consideration the above, the history and current status of local livestock breeds in Oman in general, and of chicken in particular, the aims of this thesis project were to analyze current management practices of local chickens across the major regions of the country, to describe their phenotypic and production traits, and to assess their genetic diversity using microsatellites and mtDNA. More specifically, the objectives were to:

(I)	characterize small-scale chicken production systems and management strategies in Oman's major agro-ecological zones;	Chapter 2
(II)	analyze Omani local chicken populations in terms of phenotypic diversity;	
(III)	evaluate local chickens' productive and reproductive potential under local management conditions.	
(IV)	assess the genetic variation within and between six local chicken populations using microsatellite markers;	Chapter 3
(V)	characterize the genetic structure and relatedness of local chicken populations with global reference populations (commercial and wild) at autosomal level;	
(VI)	evaluate the contribution of local populations to the total genetic diversity pool of Omani chickens for future conservation programs.	
(VII)	assess the population structure and genetic diversity of local chickens across the Arabian Peninsula at mtDNA level;	Chapter 4
(VIII)	unveil the maternal origins of chicken populations on the Arabian Peninsula;	
(IX)	determine the genetic relationships of chickens in regions sharing the Indian Ocean rim in the historic context of expansion and trading routes.	

References

AbdulNayeem M. (2000) *The rock art of Arabia: Saudi Arabia, Oman, Qatar, the Emirates and Yemen.* Hyderabad Publishers, Hyderabad, India

Aini I. (1990) Indigenous chicken production in South-East Asia. *World's Poultry Science Journal* **46**, 51–57.

Akishinonomiya F., Miyake T., Sumi S., Takada M., Ohno S. and Kondo N. (1994) One subspecies of the Red Jungle Fowl (*Gallus gallus gallus*) suffices as the matriarchic ancestor of all domestic breeds. *Proceedings of the National Academy of Sciences of the United States of America* **91**, 12505-12509.

Al-Mashakhi M. and El-Hag B. (2007) Oman, country pasture/forage resource profiles. FAO (Ed), p. 23, Rome, Italy.

Al-Saadi N. (2013) The Animal and Plant Genetic Resources Center; summary of the proposed strategy. In:. *Proceedings of the Symposium on biotechnology and conservation of species from arid regions.* SQU (Ed), Muscat, Oman, 10-13/Feb/2013.

Al-Zidjali T.M. (1996) Oman, country report to the FAO international technical conference on plant genetic resources. In: *FAO International Technical Conference on Plant Genetic Resources.* FAO (Ed), Leipzig, Germany, 17-23/Jun/1996.

Allendorf F.W. and Luikart G. (2007) *Conservation and the genetics of populations.* Blackwell Publishing, Oxford, UK.

Barua A. and Yoshimur Y. (1997) Rural poultry keeping in Bangladesh. *Journal of World's Poultry Sciences* **53**, 387-394.

Bennewitz J. and Meuwissen T.H.E. (2005) A novel method for the estimation of the relative importance of breeds in order to conserve the total genetic variance. *Genetic Selection Evaluations* **37**, 315–337.

Berthouly C., Bed'Hom B., Tixier-Boichard M., Chen C.F., Lee Y.P., Laloë D., Legros H., Verrier E. and Rognon X. (2008) Using molecular markers and multivariate methods to study the genetic diversity of local European and Asian chicken breeds. *Animal Genetics*, **39**, 121-129.

Besbes B., Tixier-Boichard M., Hoffmann I. and Jain G.L. (2007) *Future trends for poultry genetic resources.* FAO (Ed), Rome, Italy.

Beuzen N.D., Stear M.J. and Change K.C. (2000) Molecular markers and their use in animal breeding. *The Veterinary Journal* **160**, 42-52.

Biagi P. (1994) A radiocarbon chronology for the aceramic shell-middens of coastal Oman. *Arabian Archaeology and Epigraphy* **5**, 17-31.

Boettcher P.J., Ajmone-Marsan P. and Lenstra J.A. (2010) Recent development in the genetic characterization of animal genetic resources. FAO (Ed), Rome, Italy.

Boivin N., Blench R. and Fuller D.Q. (2010) Archaeological, linguistic and historical sources on ancient seafaring: a multidisciplinary approach to the study of early maritime contact and exchange in the Arabian Peninsula. In: *The evolution of human populations in Arabia: vertebrate paleobiology and paleoanthroplogy.* Petraglia M.D and Rose J.I (Eds), pp. 251-278. Springer, UK.

Bokonyi S. (1992) Preliminary information on the faunal remains from excavations at Ra's Al-Junayz (Oman). In: *Proceedings of 10th International Conference of South Asian Archaeologists in Western Europe.* Jarrige C., Gerry J.P. and Meadow R.H. (Eds). Prehistory Press, Paris, France.

Caballero A. and Toro M.A. (2002) Analysis of genetic diversity for the management of conserved subdivided populations. *Conservation Genetics*, **3**, 289-299.

CBO (Central Bank of Oman) (2011) *Annual report for the year 2010.* MoNE (Ed), Muscat, Oman.

Cheng H. and Muir M.W. (2005) Genetic selection in poultry: physiological factors associated with production and survivability. *World's Poultry Science Journal* **61**, 383-398.

Chikhi L. and Bruford M.W. (2005) Mammalian population genetics and genomics. In: *Mammalian Genomics.* Ruvinsky A. and Marshall G.J. (Eds), pp. 539-584. CABI Publishers, UK.

Cuc N.T., Simianer H., Eding H., Tieu H.V., Cuong V.C., Wollny C.B., Groeneveld L.F. and Weigend S. (2010) Assessing genetic diversity of Vietnamese local chicken breeds using microsatellites. *Animal Genetics* **41**, 545-547.

Cuc N.T.K. (2010) Vietnamese local chicken breeds: Genetic diversity and prioritising breeds for conservation. *PhD Thesis, Georg-August Universität Göttingen, Germany.*

Delany M.E. (2003) Genetic diversity and conservation of poultry. In: *Poultry genetics, breeding and biotechnology* . Muir W.E and Aggrey S.E (Eds), pp. 257-281. CABI Publishers, UK.

Desjardins P. and Morais R. (1990) Sequence and gene organization of the chicken mitochondrial genome. A novel gene order in higher vertebrates. *Journal of Molecular Biology* **212**, 599-634.

Dessie T. and Ogle B. (2001) Village poultry production systems in the Central Highlands of Ethiopia. *Tropical Animal Health and Production* **33**, 521–537.

DGALR (Direcorate General of Agriculture and Livestock Research) (2011) Agriculture and livestock, five-year research strategy (2011-2015). MAF (Ed), Muscat, Oman.

DGMAN (Direcorate General of Metorlogy and Navigation) (2012) General weather forecast for oman. PACA (Ed), Muscat, Oman.

FAO (2004) The state of the world's animal genetic resources for food and agriculture. Rischkowsky B. and Pilling D. (Eds), Rome, Italy.

FAO (2012) Progress made in stopping decline of livestock genetic diversity. http://www.fao.org/news/story/en/item/162972/icode/. *Accessed on Mar 2012.*

Fernandez J., Toro M.A. and Caballero A. (2008) Management of subdivided populations in conservation programs: development of a novel dynamic system. *Genetics* **179**, 683–692.

Frankham R., Ballou J.D. and Briscoe D.A. (2002) *Introduction to Conservation genetics.* Cambridge University, UK.

Fuller D.Q., Boivin D., Hoogervorst T. and Allaby R. (2011) Across the Indian Ocean: the prehistoric movement of plants and animals. *Antiquity* **85**, 544–558.

Fumihito A., Miyake T., Takeda M., Shingu R., Endo T., Gojobori T., Kondo N. and Ohno S. (1994) One subspecies of the red junglefowl (*Gallus gallus gallus*) suffices as the matriarchic ancestor of all domestic breeds. *Proceedings of the National Academy of Sciences of the United States of America* **91**, 12505-12509.

Galtier N., Nabholz B., Glemin S. and Hurst G.D. (2009) Mitochondrial DNA as a marker of molecular diversity: a reappraisal. *Molecular Ecology* **18**, 4541-4550.

Gandini G.C. and Oldenbroek J.K. (1999) Choosing the conservation strategy. In: *Genebanks and the Conservation of Farm Animal Genetic Resources.* Oldenbroek J.K (Ed), pp. 11–31, DLO, Lelystad, Netherland.

Gibson J., Gamage S., Hanotte O., Iñiguez L., Maillard J.C., Rischkowsky B., Semambo D. and Toll J. (2005) Options and strategies for the conservation of farm animal genetic resources. In: *Report of an International Workshop.* FAO (Ed), Montpellier, France.

Gongora J., Rawlence N.J., Mobegi V.A., Jianlin H., Alcalde J.A., Matus J.T., Hanotte O., Moran C., Austin J.J., Ulm S., Anderson A.J., Larson G. and Cooper A. (2008) Indo-European and Asian origins for Chilean and Pacific chickens revealed by mtDNA. *PNAS* **105**, 10308–10313.

Granevitze Z., Hillel J., Chen G.H., Cuc N.T., Feldman M., Eding H. and Weigend S. (2007) Genetic diversity within chicken populations from different continents and management histories. *Animal Genetics* **38**, 576-583.

Grigson C., Zarins J. and Gowlett J.A.J. (1989) Camel in Arabia: A direct radiocarbon date, calibrated to about 7000 BC. *Journal of Archaeological Science* **16**, 355–362.

Groucutt H.S. and Petraglia M.D. (2012) The prehistory of the Arabian Peninsula: deserts, dispersals, and demography. *Evolutionary Anthropology* **21**, 113–125.

Hall H. (1986) Diseases and parasites of livestock in the tropics. In: *Intermediate Tropical Agriculture Series.* Longman, UK.

Hanotte O., Bradley D.G., Ochieng J.W., Verjee Y., Hill E.W. and Rege J.E.O. (2002) African pastoralism: genetic imprints of origins and migrations. *Science* **296**, 336–339.

Hillel J., Groenen A.M., Biochard M.T., Korol A.B., David L., Kirzhner V.M., Burke T., Dirie A.B., Crooijmans R.P., Elo K., Feldman M.W., Freidlin P.J., Tanila A.M., Oortwijn M., Thomson P., Vignal A., Wimmers K. and Weigend S. (2003) Biodiversity of 52 chicken populations assessed by microsatellite typing of DNA pools. *Genetics Selection Evolution* **35**, 533-557.

Horst P., Mathur P.K. and Valle Zarate A. (1996) Breeding policies for specific tropical environments using appropriate combinations of major genes. In: *World Poultry Congress*, pp. 633-640, New Delhi, India.

IAEA (2004) Improving farmyard poultry production in Africa: Interventions and their economic assessment. In: *Proceedings of a final research coordination meeting.* Division of Nuclear Techniques in Food and Agriculture, Vienna, Austria.

Jens R., Anders P., Charlotte V., Ainsh M. and Lone F. (2004) *Keeping village poultry; A technical manual for small-scale poultry production*. 89 p. Network for Smallholder Poultry Development, Copenhagen, Denmark.

Kadim I., Al-Marzooqi W., Al-Waheebi S. and Mahgoub O. (2009) Improving egg and meat production of Omani local chicken through genetic and phenotypic selection programmes. Sultan Qaboos University, Muscat, Oman.

Leroy G., Kayang B.B., Youssao I.A., Yapi-Gnaore C.V., Osei-Amponsah R., Loukou N.E., Fotsa J.C., Benabdeljelil K., Bed'hom B., Tixier-Boichard M. and Rognon X. (2012) Gene diversity, agroecological structure and introgression patterns among village chicken populations across North, West and Central Africa. *BMC Genetics* **13**, 34.

Liu Y.P., Wu G.S., Yao Y.G., Miao Y.W., Luikart G., Baig M., Beja-Pereira A., Ding Z.L., Palanichamy M.G. and Zhang Y.P. (2006) Multiple maternal origins of chickens: out of the Asian jungles. *Molecular Phylogenetic Evolution* **38**, 12-19.

MAF (Ministry of Agriculture and Fisheries) (2013) Directorate of Rural Women; projects and programs of agricultural production and extension. Muscat, Oman.

Marle-Koster E.V. and Nel L.H. (2003) Genetic markers and their application in livestock breeding in south Africa: a review. *South African Journal of Animal Sciences* **33**, 1-10.

Mcainsh C., Kusina J., Madsen J. and Nyoni O. (2004) Traditional chicken production in Zimbabwe. *World's Poultry Sciences Journal* **60**, 233-246.

Mendelsohn R. (2003) The challenge of conserving indigenous domesticated animals. *Ecological Economics* **45**, 501-510.

Miao Y.W., Peng M.S., Wu G.S., Ouyang Y.N., Yang Z.Y., Yu N., Liang J.P., Pianchou G., Beja-Pereira A., Mitra B., Palanichamy M.G., Baig M., Chaudhuri T.K., Shen Y.Y., Kong Q.P., Murphy R.W., Yao Y.G. and Zhang Y.P. (2013) Chicken domestication: an updated perspective based on mitochondrial genomes. *Heredity* **110**, 277–282.

MoNE (Ministry of National Economy) (1995) The Vision for Oman's Economy 2020. Muscat, Oman.

Mtileni B.J., Muchadeyi F.C., Maiwashe A., Chimonyo M., Groeneveld E., Weigend S. and Dzama K. (2011a) Diversity and origin of South African chickens. *Poultry Sciences* **90**, 2189–2194.

Mtileni B.J., Muchadeyi F.C., Maiwashe A., Groeneveld E., Groeneveld L.F., Dzama K. and Weigend S. (2011b) Genetic diversity and conservation of South African indigenous chicken populations. *Journal of Animal Breeding and Genetics* **128**, 209-218.

Muchadeyi F.C., Eding H., Simianer H., Wollny C.B.A., Groeneveld E. and Weigend S. (2008) Mitochondrial DNA D-loop sequences suggest a Southeast Asian and Indian origin of Zimbabwean village chickens. *Animal Genetics* **39**, 615–622.

Muchadeyi F.C., Eding H., Wollny C.B., Groeneveld E., Makuza S.M., Shamseldin R., Simianer H. and Weigend S. (2007) Absence of population substructuring in Zimbabwe chicken ecotypes inferred using microsatellite analysis. *Animal Genetics* **38**, 332-339.

Mwacharo J.M., Bjørnstad G., Han J.L. and Hanotte O. (2013_a) The history of African village chickens: an archaeological and molecular perspective. *African Archaeological Review* **30**, 97–114.

Mwacharo J.M., Bjornstad G., Mobegi V., Nomura K., Hanada H., Amano T., Jianlin H. and Hanotte O. (2011) Mitochondrial DNA reveals multiple introductions of domestic chicken in East Africa. *Molecular Phylogenetic Evolution* **58**, 374-382.

Mwacharo J.M., Nomura K., Hanada H., Han J.L., Amano T. and Hanotte O. (2013_b) Reconstructing the origin and dispersal patterns of village chickens across East Africa: insights from autosomal markers. *Molecular Ecology* **22**, 2683–2697.

Mwacharo J.M., Nomura K., Hanada H., Jianlin H., Hanotte O. and Amano T. (2007) Genetic relationships among Kenyan and other East African indigenous chickens. *Animal Genetics* **38**, 485–490.

Mwalusanya N., Katule A., Mutayoba S., Mtambo M., Olsenand J. and Minga U. (2002) Productivity of local chicken under village management conditions. *Tropical Animal Health and Production* **34**, 405–416.

Nei M. and Li W.H. (1979) Mathematical model for studying genetic variation in terms of restriction endonucleases. *PNAS* **76**, 5269–5273.

Nishibori M., Shimogiri T., Hayashi T. and Yasue H. (2005) Molecular evidence for hybridization of species in the genus Gallus except for Gallus varius. *Animal Genetics*, 367–375.

Notter D.R. (1999) The importance of genetic diversity in livestock populations of the future. *Journal of Animal Sciences* 77, 61-69.

Oka T., Ino Y., Nomura K., Kawashima S., Kuwayama T., Hanada H., Amano T., Takada M., Takahata N., Hayashi Y. and Akishinonomiya F. (2007) Analysis of mtDNA sequences shows Japanese native chickens have multiple origins. *Animal Genetics* **38**, 287-293.

Ollivier L. and Foulley J.L. (2005) Aggregate diversity: new approach combining within- and between-breed genetic diversity. *Livestock Production Science* **95**, 247–254.

Perera O. (2010) The changing pattern of livestock farming systems in South-Asia, threats to indigenous breeds, and programs to conserve animal genetic resources. In: *Proceedings of the Workshop Sponsored by OECD on International Strategic Programs for the Conservation of Animal Genetic Resources for Food and Agriculture*. Lessard C. (Ed), 17 July 2010, Vancouver, Canada.

Perez-Figueroa A., Saura M., Fernandez J., Toro M.A. and Caballero A. (2009) METAPOP—a software for the management and analysis of subdivided populations in conservation programs. *Conservation Genetics* **10**, 1097–1099.

Pica-Ciamarra U. and Dhawan M. (2010) Small-scale poultry farming and poverty reduction in South Asia: from good practices to good policies in Bangladesh, Bhutan and India. In: *South-Asia Pro-Poor Livestock Policy Programme*. NDDB and FAO (Eds).

Pickering H. (2007) *Frankincense; the perfume of Arabia*. Al Roya Press and Publishing House, Muscat, Sultanate of Oman.

Razafindraibe H., Mobegi V.A., Ommeh S.C., Rakotondravao M.L., Bjornstad G., Hanotte O. and Jianlin H. (2008) Mitochondrial DNA origin of indigenous Malagasy chicken. *Annals of the New York Academy of Sciences* **1149**, 77-79.

Reynolds J., Weir B.S. and Cockerham C. (1983) Estimation of the coancestry coefficient: Basis for a short-term genetic distance. *Genetics* **105**, 767-779.

Romanov M.N. and Weigend S. (2001) Analysis of genetic relationships between various populations of domestic and red jungle fowl using microsatellite markers. *Poultry Science* **80**, 1057-1063.

Saleh F. (2000) Improving the local Dhofari poultry. Ministry of Agriculture and Fisheries, Dhofar, Sultanate of Oman.

Saura M., Perez-Figueroa A., Fernandez J., Toro M.A. and Caballero A. (2008) Preserving population allele frequencies in ex situ conservation programs. *Conservation Biology* **22**, 1277-1287.

SCTA (Saudi Commission of Tourism and Antiquities) (2013) The Roots of Purebred Arabian Horses. *http://www.scta.gov.sa/en. Accessed in Aug 2013.*

Shahbazi S., Mirhosseini S.Z. and Romanov M. (2007) Genetic diversity in five Iranian native chicken populations estimated by microsatellite markers. *Biochemical Genetics* **45**, 63–75.

West B. and Zhou B.X. (1988) Did chickens go north? New evidence for domestication. *Journal of Archaeological Science* **15**, 515-533.

Wilkens B. (2005) Fishing in the Arabian Sea: A short note on the prehistoric sites Ras Al-Had and Ra's Al-Jinz in Oman. *Paleorient* **31**, 126-130.

Wollny C.B.A. (2003) The need to conserve farm animal genetic resources in Africa: Should policy marker be concerned? *Ecological Economics* **45**, 341-351.

Zanetti E. (2009) Genetic, phenotypic and proteomic characterisation of local chicken breeds. *PhD Thesis, Université de Toulouse, Italy.*

Zeder M.A., Emshwiller E., Smith B.D. and Bradley D.G. (2006) Documenting domestication: the intersection of genetics and archaeology. *Trends in Genetics* **22**, 139-155.

Supplementary materials

Figure S1.1 Cave drawing showing fat-tailed sheep on a rock west of Bahla. Source: AbdulNayeem (2000).

Figure S1.2 Small-scale local chicken unit provided by the Ministry of Agriculture and Fisheries. Source:MAF (2013)

CHAPTER 2

Towards conservation of Omani local chicken: Phenotypic characteristics, management practices and performance traits

B. Al-Qamashoui[1], O. Mahgoub[2], I. Kadim[2] and E. Schlecht[1]

[1] *Animal Husbandry in the Tropics and Subtropics, University of Kassel and Georg-August-Universität Göttingen, Albrecht-Thaer-Weg 3, 37075 Göttingen, Germany.*

[2] *Department of Animal and Veterinarian Sciences, Sultan Qaboos University, Po.Box 50, 123 Muscat, Sultanate of Oman.*

Published in a slightly modified version in:

Asian-Australasian Journal of Animal Sciences

Vol. 27, No. 6: 767-777, June 2014

Abstract

Characterizing local chicken types and their mostly rural production systems is prerequisite for designing and implementing development and conservation programs. This study evaluated the management practices of small-scale chicken keepers and the phenotypic and production traits of their chicken in Oman, where conservation programs for local livestock breeds are currently started. Free-range scavenging was the dominant production system, and logistic regression analysis showed that socio-economic factors such as training in poultry keeping, household income, income from farming and gender of chicken owners influenced feeding, housing, and health care practices ($P<0.05$). A large variation in plumage and shank colors, comb types and other phenotypic traits within and between Omani chicken populations were observed. Male and female body weight differed ($P<0.05$), being 1.3 ± 0.65 kg and 1.1 ± 0.86 kg respectively. Flock size averaged 22 ± 7.7 birds per household with 4.8 hens per cock. Clutch size was 12.3 ± 2.85 and annual production 64.5 ± 2.85 eggs per hen. Egg hatchability averaged 88 ± 6.0% and annual chicken mortality across all age and sex categories was 16 ± 1.4%. The strong involvement of women in chicken keeping makes them key stakeholders in future development and conservation programs, but the latter should be preceded by a comprehensive study of the genetic diversity of the Omani chicken populations.

Keywords: Animal genetic resources; egg production; rural smallholders; scavenging system; task division.

2.1 Introduction

Local chickens play an important role for smallholders and contribute significantly to food security of households in rural and semi-urban communities (Abdelqader et al. 2007). According to Jens et al. (2004), nearly all rural and semi-urban families in developing countries keep a small flock of local chickens in the backyard. Scavenging systems and low input into feeding, housing and labor as well as adaptation to diseases, absence of veterinary services and poor management (Hall 1986) are considered as the main characteristics of local chicken production systems in tropical and subtropical countries (Aini 1990; Gueye 2000). A considerable phenotypic variation is another main characteristic of local chicken types throughout the world (Mcainsh et al. 2004). Women are frequently in charge of local chicken husbandry (Mwalusanya et al. 2002) and are especially involved in most activities of poultry management, although a division of labor often exists within the household (Kondombo et al. 2003). However, rural communities often lack the required husbandry skills, training and market opportunities to effectively improve their chicken production (Mwalusanya et al. 2002).

In Oman where more than 40% of the population is still engaged in the agricultural sector (MoNE 2010), no studies have been carried out so far to characterize and develop the rural chicken production systems for conservation purposes. Since the design of conservation and development programs requires full characterization of village production systems (Gueye 2000), the current study aimed at analyzing (1) Omani rural chicken populations in terms of phenotypic diversity; (2) small-scale chicken production systems and marketing strategies in Oman's major agro-ecological zones; (3) local chicken's productive and reproductive potential under different management conditions; and (4) overall opportunities and constraints of traditional small-scale chicken farming in Oman.

2.2 Materials and methods

2.2.1 Study locations, interviews and data collection

The study was carried out in the six major agro-ecological zones (AEZ) of Oman, namely Musandam (MU), Batinah (BT), North Hajar (NH), East Hajar (EH), East Coast (EC), and Dhofar (DF). These zones (Figure 2.1) are clearly apart from each other and differ widely in topographic aspects, climate (Table 2.1), soils and agricultural production systems (DGALR 2011).

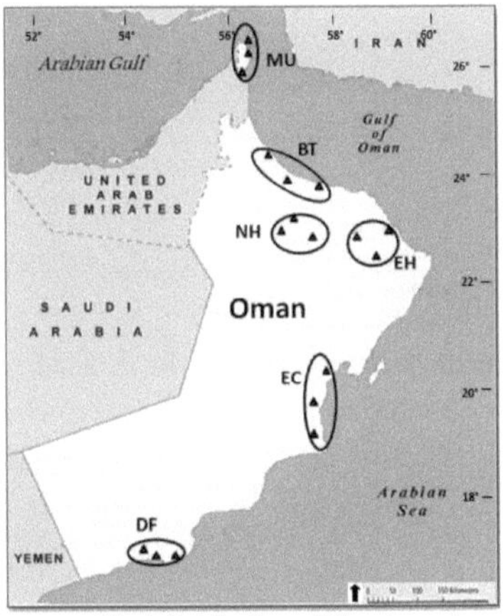

Figure 2.1 Oman map showing the geographical distribution of the six major agro-ecological zones (AEZ) in circles and sampling areas within each zone in triangles (Source: MoNE (2010)). See above text for abbreviations of AEZ names.

Three villages were selected from each AEZ according to the information given by the regional Agricultural Directorates. In cooperation with agents of the local agricultural extension centers, a preliminary survey was conducted to gather principal information concerning small-scale farmers in the six AEZ.

Table 2.1 Climatic and topographic features and main agricultural activities in six major agro-ecological zones (AEZ) of Oman.

AEZ	Average temperature (°C) Cool season	Average temperature (°C) Hot season	Yearly average humidity (%)	Rainfall (mm/ year)	Topographic features	Farming activities
Musandam (MU)	16	40	90	192	A peninsula of steep rocks.	Livestock rearing. Little cultivation.
Batinah (BT)	15	35	63	99	Long, narrow flat coastal strip; fertile plain.	Date palm, fruits, vegetables and crop cultivation. Livestock rearing.
North Hajar (NH)	14	35	25	345	Mostly steep mountains; highest and wildest terrain in the country.	Fruit and crop cultivation, date palm. Livestock rearing.
East Hajar (EH)	15	34	70	30	Mountains, midland and lowland; mostly steep and barren formations of igneous and sedimentary rocks; very dry.	Livestock rearing. Fruit and crop cultivation, date palm.
East Coast (EC)	18	34	80	67	Sandy coast and inlands.	Livestock. No cultivation.
Dhofar (DF)	18	32	88	200	Mountains, midland and plain; tropical climate through most of the year, influenced by monsoon in summer.	Coconut, fruits, vegetables, annual grass cultivation, Livestock.

Sources: DGALR (2011); MoNE (2010)

A total of 163 households were selected for the detailed study (20 - 30 households from each AEZ, distributed across 3 villages) using a stratified sampling method. In each AEZ the selected farms had similar agricultural systems and were representative for the zone. Villages in close proximity to large cities were avoided.

The households in the study villages were visited and data were collected using a pre-tested structured questionnaire covering households' socio-demographic and economic characteristics, and characteristics of their livestock and cropping activities in general. Number of chicken, egg production, health care, feeding and housing strategies, bird ownership as well as decision-making were recorded. Normally the head of the family (householder) or flock caretaker was interviewed once during the study period. However, in some cases the visit was repeated on selling and purchasing days of new chicken stocks or when new houses for the chicken were built.

2.2.2 Measuring morphological traits of chicken

A total of 199 adult chickens aged 9 to 12 months were selected for the assessment of phenotypic traits according to the following distribution: DF - 20 females, 6 males; EC - 25 females, 6 males; EH - 28 females; 6 males; MU - 30 females, 6 males; NH - 30 females, 6 males; BT - 30 females, 6 males. Variables measured included body weight, body length (distance from the beginning of the neck to the tail) and shank length (length of the tarsometatarsus from the hock joint to the metatarsal pad). Body and shank lengths were measured using a graduated tape while the bird was standing upright; body weight was measured in kilogram using an electronic hanging scale (accuracy 0.01 g). The recorded morphological traits included plumage, eye, comb, and shank colors and patterns. Data collection was completed by taking a picture of each surveyed bird.

2.2.3 Statistical analysis

Descriptive statistics on the phenotypic traits were computed using SAS Version 9.3 (SAS Institute Inc., Cary, NC). For the management part, data analysis was performed using SPSS 19.0 (SPSS Inc., an IBM Company, Chicago, IL). Differences between AEZ were explored using Chi-square test (categorical variables) or Kruskal-Wallis test (continuous variables), whereby continuous variables were first tested for normality (Kolmogorov-Smirnov test). Kendall's coefficient of concordance was computed to assess the major mean ranks of chicken traits given priority by flock owners when selecting new chicken flocks. The major farming activities of households were assessed using weighted means procedures, with each activity being weighted according to its order among the three first important activities.

A stepwise logistic regression with backward elimination of predictors (Hair et al. 2006) was used to relate chicken keepers' adoption of supplementary feeding of birds (yes/no) and of solid housing (yes/no) to independent predictors. Several independent variables (among others, AEZ, age of householder, total household income, farm contribution to total income, cropland size, chicken flock size, years of experience in chicken keeping, training in poultry keeping) were included in the full model [Eq. 1]:

$$\text{Logit}\ (Y_{1/0}) = \beta_0 + \beta_1 X_1 + \beta_2 X_2 + \ldots + \beta_n X_n + \varepsilon \qquad \text{[Eq. 1]}$$

where Y is the dependent variable, and $\beta' = (\beta_0, \beta_1 \ldots)$ the model parameters to be estimated, ε the error term and X_i the independent variables. The fit of the final model was assessed by the model Chi-square (Model X^2) and the Hosmer and Lemeshow goodness-of-fit test (Archer and Lemeshow 2006). Well-fitting models showed significance ($P<0.05$) on the Model X^2 and non-significance ($P>0.05$) on the goodness-of-fit test.

A multiple linear regression (Eq. 2) was used to predict chicken flock size, egg production, and bird survival rate from different socio-economic and management variables (among others, family size, gender and age of chicken owner, years of experience in chicken keeping, daily scavenging period, offer of commercial feeds, equipment use, presence of a solid chicken house, presence of hired labor, cleaning of chicken house and utensils, administration of medicine) as follows:

$$Y_i = a + b_1 X_1 + b_2 X_2 + \ldots + b_k X_k + \varepsilon \qquad \text{[Eq. 2]}$$

where Y is the dependent variable, a the intercept, b_i the regression coefficient, ε the error term and X_i the predictor variable.

2.3 Results and discussion

2.3.1 Household socioeconomic characteristics and farming activities

Of the interviewed 163 householders, 125 (76.8%) were male and 38 (23.2%) female (Table 2.2). On a weighted means basis, date palm cultivation, small ruminant husbandry, fruit and vegetable cropping and cereal and fodder cultivation were the most important farming activities in BT, EH, NH and MU (Figure 2.2). In DF, the major farming activity was fruit and vegetable cropping, while in EC small ruminant husbandry was dominant. Chicken husbandry is a widely spread activity of rural smallholder farmers across Oman even though, from an economic point of view, its importance is inferior to that of the production of dates, cereals and fodder crops, fruits and vegetables and ruminant livestock (Figure 2.2).

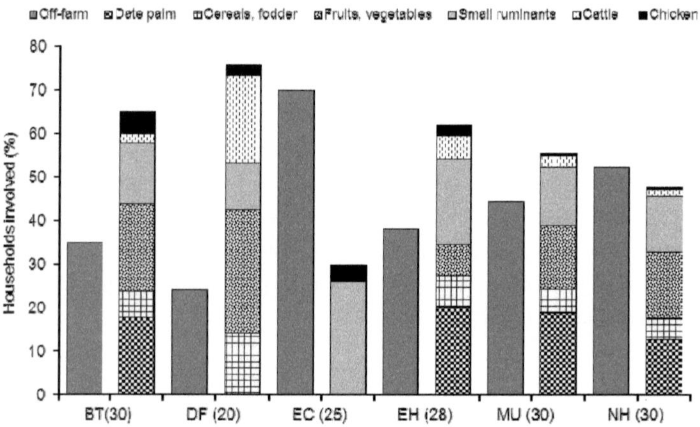

Figure 2.2 Off-farm engagement and major agricultural activities of 163 smallholder farmers across six agro-ecological zones (AEZ) of Oman as derived from weighted means computation. Figures in parenthesis depict the number of interviewed households per zone. See Materials and methods section 2.2.1 for AEZ name abbreviations.

For 68.9% of the respondents the main reason for keeping chicken was home consumption of eggs and meat, whereas 31.3% reported to sell some of the live chicken and eggs (Table 2.2). However, the exact contribution of chicken to household income and self-sufficiency in poultry meat and eggs could not be determined, partly due to lack of reliable production data and recalls.

2.3.2 Ownership and task division in chicken farming

Although the householders across the studied regions were mostly men, chicken ownership was dominated by females in all AEZ; they controlled the inflow and outflow of birds and were involved in selling and selecting new flocks. Within the family, chickens were primarily owned by women aged 15 - 60 years (70.8%; Table 2.2). Women and children below 15 years of age were highly involved in daily chicken management, especially feeding, watering and egg collection (58.7%; Table 2.2). External (male) laborers and husbands were primarily responsible for the maintenance of the chicken houses and equipment. The selection of birds among growing chicks and the purchasing of new birds for breeding or replacement was the task of women (85.9%).

Table 2.2 Household characteristics, ownership patterns, and responsibilities for and purpose of keeping local chicken flocks by 163 smallholder farmers across six agro-ecological zones (AEZ) of Oman. All values are percentages of occurrence in the different zones and across the zones (last column). Sums of percentages per category can deviate from 100. See Materials and methods section 2.2.1 for AEZ name abbreviations.

AEZ Variables	MU *n*=30	BT *n*=30	NH *n*=30	EH *n*=28	EC *n*=25	DF *n*=20	Mean
Household head, sex							
Female	16.7	33.3	26.6	21.4	16.0	25.0	23.2
Male	83.3	66.7	73.4	78.6	84.0	75.0	76.8
Age/sex group of chicken owners							
Children (<15 yrs)	3.3	3.3	6.4	9.9	7.5	0.0	5.1
Male (15 – 60 yrs)	13.5	8.9	10.7	17.8	16.1	10.8	13.0
Female (15 – 60 yrs)	74.4	77.6	69.0	60.7	64.0	79.2	70.8
Older members (>60 yrs)	8.8	10.2	13.9	11.6	12.4	10.0	11.2
Source of specific knowledge of chicken owner							
Traditional knowledge	62.1	46.7	56.7	69.7	82.0	65.9	63.8
Technical training	37.9	53.3	43.3	30.3	18.0	34.1	36.2
Responsible for feeding, watering, cleaning and collecting eggs							
Wives and children	64.7	44.2	64.4	69.6	70.7	38.5	58.7
Husband	17.8	17.8	22.8	15.3	17.0	20.0	18.5
External labor	17.4	37.8	12.8	15.1	12.3	41.7	22.9
Responsible for maintenance of chicken houses and assets							
Husbands	27.8	14.9	33.8	29.1	32.9	11.0	23.4
External labor	72.2	85.1	66.2	70.1	67.1	89.0	76.6
Responsible for selecting and purchasing birds, and selling products							
Wives	91.1	83.3	89.9	84.4	95.3	71.1	85.9
Husband	8.9	16.7	10.1	15.6	14.7	28.9	15.8
Purpose of keeping local chicken							
Home consumption and income	36.7	23.3	36.7	24.1	41.6	25.5	31.3
Home consumption only	63.3	76.7	63.3	75.9	59.4	74.5	68.9

Table 2.3 Mean rank[1] and placement[2] of criteria for the selection of replacement chickens provided by 163 smallholder farmers across six agro-ecological zones (AEZ) of Oman. See Materials and methods section 2.2.1 for AEZ name abbreviations.

AEZ Major selection traits	MU *n*=30	BT *n*=30	NH *n*=30	EH *n*=28	EC *n*=25	DF *n*=20
Egg production	2.5 [1]	2.3 [1]	2.1 [1]	2.3 [1]	2.2 [1]	2.2 [1]
Egg size	3.3 [4]	3.5 [5]	3.9 [5]	3.5 [5]	3.7 [5]	3.5 [5]
Body size and growth rate	2.8 [2]	2.5 [2]	2.3 [2]	2.5 [2]	2.5 [2]	2.8 [2]
Body conformation	3.4 [5]	3.3 [3]	3.4 [4]	3.3 [3]	3.4 [4]	3.4 [4]
Feather color	3.0 [3]	3.4 [4]	3.3 [3]	3.4 [4]	3.3 [3]	3.0 [3]
W[3]	0.07	0.20	0.34	0.20	0.24	0.14

[1] Mean rank: 1 highest, 5 lowest.
[2] Placement (rank) across all variables per zone given in square brackets.
[3] W: Kendall's Coefficient of Concordance (0 = no agreement, 1 = total agreement).

As far as specific selection criteria for new chicken were concerned, all respondents selected replacement chickens based on one or more criteria, in particular egg production, egg size, body size and growth rate of the mother, body conformation and feather color (Table 2.3). When selecting hatching eggs, farmers declared that eggs for the next generation should be collected from hens with a good performance history.

2.3.3 Housing and feeding management

The respondents used different locally available and cheap building materials for constructing chicken houses; these only hosted birds of the own family. Wood sticks and sheets, palm leafs, fabric and corrugated iron were the main materials used. Solid concrete and stone houses were relatively frequent in MU and BT (Table 2.4). Light was hardly available in the chicken houses, however, electrical pear lamps for brooding were frequent in DF (75%),while fans for air circulation were very rarely used across all AEZ (7%). Preparing nesting boxes for the hens was common in all AEZ but least frequent in EC. Nests were made from cheap local materials such as a large tin with cut ends, or wood.

During daytime, birds were released to scavenge freely on agricultural by-products, household wastes, in the fields or home gardens or close to their shelters. During night they were confined in their houses. However, commercial supplements (mainly feed concentrates) were additionally given to the birds by 58.5% of the respondents. The scavenging system with the use of household wastes and plant by-products was also reported from Malawi (Gondwe 2004), Ethiopia (Dessie and Ogle 2001) and Burkina Faso (Kondombo et al. 2003). However, the nutrient values of such scavenged by-products and wastes need to be evaluated. Abdelqader et al. (2007) suggested that meeting the nutrient requirements of scavenging chicken depends on the available scavenging area per bird, the quality of scavenging feed resources, the season and the birds' production stage.

Table 2.4 Construction material for chicken houses, housing equipment and feeding system used by 163 smallholder farmers across six agro-ecological zones (AEZ) of Oman. All values are percentages of use by farmers in the different zones and across the zones (last column). Sums of percentages per category can deviate from 100. See Materials and methods section 2.2.1 for AEZ name abbreviations.

AEZ Variable	MU *n*=30	BT *n*=30	NH *n*=30	EH *n*=28	EC *n*=25	DF *n*=20	Mean
Construction material							
Wooden and iron sheet	50.1	50.7	66.7	67.8	92.0	69.2	66.1
Concrete/ mud	29.9	39.4	23.3	21.4	4.0	20.4	23.1
Palm leaves and fences	20.0	9.9	10.0	10.8	4.0	10.4	10.9
Existence of management assets							
Brooding lamp	20.0	26.7	16.2	25.0	0.0	75.0	27.2
Laying nests	43.3	53.3	43.3	28.6	8.7	39.8	36.1
Air circulation fans	8.0	10.0	10.1	7.1	0.0	5.2	6.7
Approved feeders and water roughs	36.6	71.9	58.0	22.2	20.3	75.9	47.5
Feeding system							
Scavenging only	33.5	26.7	33.3	47.1	53.5	55.0	41.5
Use of commercial supplements	66.5	73.3	66.7	52.9	46.5	45.0	58.5

The level at which farmers maintain their bird flocks was of interest in our study. A binary logistic regression was employed to investigate farmers' decision to house their flock in solid houses and to feed them a commercial supplemental feed (Table 2.5). Training in poultry husbandry, cropland area, contribution of farm income to household income and flock size showed a significant ($P<0.05$) and positive correlation with keeping the birds in solid houses. Training in poultry keeping and a higher total household income increased the likelihood of offering supplement feeds to chicken ($P<0.05$). Resource availability might at least partly have influenced the type of housing structures chosen by the famers (Ramlah 1996). For Botswana, Badubi et al. (2006) reported that good housing improved flock productivity in free-range scavenging systems. The significant effect of flock size indicated that farmers provided better protection from predators and environmental conditions when chicken numbers increased. Yet, income from farming and training in poultry keeping were the strongest predictors for improved housing.

Table 2.5 Coefficients of the logistic regression models predicting the decision of 163 smallholder farmers to keep local chickens in solid houses (above) and to offer purchased supplementary feed (below) across six major agro-ecological zones of Oman.

Regression parameters	ß	$SE_ß$	Wald's χ^2	df	$P\leq$	Odds ratio
Dependent variable: Keep chicken in solid house *(yes)*						
Constant	−16.35	3.85	18.02	1	0.001	n.a.
Training in poultry keeping (yes = 1)	3.91	1.16	11.29	1	0.001	49.72
Cropland size (feddan)[1]	1.44	0.51	7.86	1	0.005	4.20
Farming contributes to income (yes = 1)	4.47	1.40	10.16	1	0.001	87.18
Chicken flock size (n)	0.29	0.09	10.18	1	0.001	1.33
Overall model evaluation (Model X^2)			152.44	4	0.001	
Goodness-of-fit test[2]			45.27	8	0.691	
Dependent variable: Offer commercial supplement feeds *(yes)*						
Constant	−10.92	1.92	30.89	1	0.001	n.a.
Total income of household (OMR/yr)[1]	0.02	0.01	30.20	1	0.001	1.02
Training in poultry keeping (yes = 1)	3.94	1.29	9.22	1	0.002	51.28
Overall model evaluation (Model X^2)			155.94	2	0.001	
Goodness-of-fit test[2]			5.43	6	0.49	

[1] Units: feddan = Arabic unit of area, 4200 m²; OMR = Omani Rial, exchange rate 1 OMR = 2.6 USD

[2] Hosmer and Lemeshow Goodness-of-fit test (Archer and Lemeshow 2006).

n.a.= Not applicable; for binary variables, yes = 1 and no = 0.

Approximately 36% of the interviewees benefited from technical services provided by extension agents or veterinarians, or had received advice and technical training in poultry management (results not shown). Training in poultry husbandry by extension agents increased the farmers' likelihood to offer commercial supplement feed to their birds, pointing to the effectiveness of extension programs in improving the productivity of the chicken business. Adebayo and Adeola (2005) indicated that the relationship between skill level and flock production is directly linked to the level of knowledge and management, which contribute to the profitability of their business.

2.3.4 Phenotypic characteristics and production traits of chicken

Local chicken were mostly normally feathered (hens 68.1%, cocks 83.3%) with a few showing soft and fluffy feathers (hens 23.9%, cocks 16.7%). Very diverse plumage coloration of neck, breast and wing was observed (Table S2.1), with pale brown (27%), deep dark brown (27%) and deep dark brown (26.4%), respectively, being the dominant color for these areas in hens. Neck, breast and wing plumage in cocks were predominantly colored in shining orange-yellow (58.3%), black (44.4%) and shining orange-yellow (36.1%), respectively. Most chicken showed very light skin color (hens 75.5%, cocks 38.9%), whereas dark colored skin existed in 21.5% of hens and yellow and very dark skin were observed at 30.6% each in cocks. The predominant beak color was yellow (hens 64.4%, cocks 41.7%), followed by black to very dark (hens 25.2%, cocks 36.1%) and beige to brown (hens 8.0%, cocks 22.2%). The commonest comb color was red (hens 77.9%, cocks 83.3%), while 4.3% of hens and 16.7% of cocks showed black to very dark red/blue colors. A significant domination ($P<0.05$) of the single comb in females (74.2%) and males (66.7%) was observed. The predominant iris color was orange/red (hens 74.2%, cocks 55.6%) followed by brown/black (hens 23.9%, cocks 38.9%) and white/yellow (hens 1.8%, cocks 5.6%). The shank color varied between blue-gray (40.5%), white (33.1%), yellow (16.0%) and black (9.2%) in females, and between yellow (36.1%), blue-gray (27.8%), black (25.0%) and white (11.1%) in males.

The large variation in plumage color might be attributed to a lack of selection of breeders for this trait, which was also reported from Nigeria (Daikwo et al. 2011), Jordan (Abdelqader et al. 2007) and Botswana (Badubi et al. 2006). Fisseha (2009) suggested that the presence of such large variation in color of plumage and other morphological attributes of chicken ecotypes within regions may be the result of the absence of geographical isolation as well as long periods of natural selection. Light/pink skin and red comb color in females and males dominated in all our study zones, which agrees with the findings of Barua and Yoshimur (1997) for local chicken in Bangladesh. The light color of comb and skin might contribute to the birds' tolerance of heat stress (Van Kampen 1974; Egahi et al. 2010). From the analysis of 29 autosomal markers it appears that two subspecies of red jungle fowl, namely *Gallus gallus gallus* from Thailand and *Gallus gallus spadicus* from China, are quite distant from Omani chicken (Al-Qamashoui et al. 2014a), while analysis of mtDNA indicated that Indian chicken, including subspecies *Gallus gallus murghi*, seem to be more closely related to the local populations of Omani chicken (Al-Qamashoui et al. 2014b), which can be explained by the historically very intense trade of seafarers from the Arabian Peninsula with the Middle East and Indian region (Biagi 2006; Boivin and Fuller 2009).

The mean body weight of local cocks and hens across Oman (1.24 kg) is similar to values from Namibia (Petrus et al. 2011) and central Nigeria (Daikwo et al. 2011), while higher weights were reported from Jordan (Abdelqader et al. 2007) and Botswana (Badubi et al. 2006). At 1.33 ± 0.65 kg, the mean body weight (Table 2.6) of adult cocks was significantly ($P<0.05$) heavier than that of hens (1.17 ± 0.86 kg). Cocks also had higher values ($P<0.05$) for body length (18.4 ± 0.14

cm) and shank length (8.1 ± 0.11 cm) than hens (17.3 ± 0.13 cm; 7.1 ± 0.14 cm). While clutch size was not related to body length and shank length of hens (r<0.4, *P*>0.05), there was a significant correlation between body weight and clutch size (r=0.66, *P*<0.05). The differences in body weight and body measures between male and female birds are in agreement with reports from Tanzania (Mwalusanya et al. 2002) and Zimbabwe (Mcainsh et al. 2004); such differences are due to the differential effects of androgens and estrogens on growth (Yakubu et al. 2009). The higher body weight of male and female chickens in DF than in the other AEZ might be attributed to less efforts needed by these birds to scavenge their feed: DF farms are smaller-sized than farms in the other AEZ but characterized by highly productive vegetable cultivation, potentially offering plenty of nutritious residues.

Table 2.6 Body weight, body and shank lengths (Means* ±SD) of 199 local chicken across six major aagro-ecological zones (AEZ) of Oman. See Materials and methods section 2.2.1 for AEZ name abbreviations.

AEZ	Birds (n)		Body weight (kg)		Body length (cm)		Shank length (cm)	
	Male	Female	Male	Female	Male	Female	Male	Female
MU	6	30	1.4 ± 0.29	1.1 ± 0.09	18.3 ± 0.11	17.5 ± 0.13	8.5 ± 0.22	6.9 ± 0.13
BT	6	30	1.3 ± 0.15	1.2 ± 0.10	18.7 ± 0.21	17.5 ± 0.09	8.2 ± 0.17	6.9 ± 0.14
NH	6	30	1.3 ± 0.42	1.2 ± 0.11	18.5 ± 0.19	17.0 ± 0.13	8.5 ± 0.22	6.9 ± 0.14
EH	6	28	1.2^{a} ± 0.41	1.0 ± 0.09	18.3 ± 0.17	17.6 ± 0.20	8.3 ± 0.21	6.8 ± 0.16
EC	6	25	1.4 ± 0.37	1.1 ± 0.10	17.2 ± 0.17	16.8^{a} ± 0.11	7.7 ± 0.33	7.2 ± 0.13
DF	6	20	1.4 ± 0.14	1.4^{a} ± 0.28	18.8 ± 0.17	18.2 ± 0.14	8.0 ± 0.36	8.1^{a} ± 0.16

* Within columns (i.e., between AEZ) values with a superscript differ at *P*<0.05 from the others (Kruskal-Wallis test).

Age at sexual maturity of the hen, defined as age when producing the first egg, was reported to be 24.1 ± 1.33 weeks (Table 2.7), occurring earlier in BT (20.7 ± 1.29) and DF (20.0 ± 1.80) than in the other AEZ (*P*<0.05). Omani hens were maturing at the same pace as hens in Ethiopia (6.7 months; Dessie and Ogle 2001), and Malawi (6.1 months; Gondwe 2004). The hens produced on average 5.2 ± 0.23 clutches per year with a total of 12.3 ± 2.85 eggs per clutch (range 8 - 14), resulting in 64.5 ± 6.91 eggs per hen and year. The latter value was higher than that reported for local chicken in Bangladesh (44; Baru and Yoshimur 1997) and Uganda (40-50; Ssewannyana et al. 2008), while it was similar to the production reported from Tanzania (Mwalusanya et al. 2002) and Botswana (Badubi et al. 2006). The proportion of hatched eggs per clutch was 88.1 ± 6.01% with significant differences between EH (92.9 ± 7.16) and the other AEZ (*P*<0.05). The egg hatchability across Omani smallholder systems is within the range reported from Burkina Faso (60 – 90%; Kondombo et al. 2003) and higher than values reported from Botswana (42%; Badubi et al. 2006) and Nigeria (48%; Daikwo et al. 2011). Hatchability of eggs depends on hygienic and incubation conditions in the nests, egg quality, nutrition of the breeding hen, genetic factors and diseases (Sainsbury 1992). In our study, the high hatchability might be partly attributed to the high number of breeding cocks per flock. The results of the multiple linear regression analysis (Table 2.8) indicated that total egg production was significantly (*P*<0.05) higher with increasing years of experience of the chicken owner, old age of the householder and the daily frequency of supplement feeding. In addition to the positive effect of better nutrition on chicken performance, feeding chicken several times a day allows the farmer to observe the flock and notice any problem. Since a quantification of chickens' daily feed intake was not feasible in the context of the present study, it was also not possible to relate the observed variation in body conformation and production traits to differences in feeding management.

Yearly bird mortality (total number of birds that died divided by average yearly flock size) was 16.4 ± 1.37% with the highest percentage ($P<0.05$) reported from DF (17.0 ± 1.21%). Lack of adequate housing can partly explain the mortality, as good housing is a prerequisite for any viable and sustainable chicken operation (Fisseha 2009). The multiple linear regression analysis indicated that the yearly survival rate of the chicken depended on the provision of medicine and health treatments to the chicken, and was in addition positively affected by hiring external labor, but negatively related to old age of the householder (Table 2.8). The latter seems to indicate that management intensity declines with advanced age of the farmer, which might be due to poor willingness of elderly persons to take risk in overall farm management (Mandleni and Anim 2012).

Average flock size across all AEZ, calculated as mean of the current size and the maximum and minimum flock size during the past 10 years, was 21.9 ± 7.69 birds and varied between 12 and 41 (Table 2.7). Flock size in EC (14.6 ± 2.10) was lowest ($P<0.05$) whereas it was highest in BT (28.7 ± 7.65). At least one cock was kept in each flock for breeding purposes. The average sex ratio was 2.1 ± 0.92 cocks per 10 females. The present chicken flock size was in the range of values reported from northern Ethiopia (12; Fisseha 2009), and Uganda (18; Ssewannyana et al. 2008). Larger flock sizes were reported from Mauritius (60; Jugessur et al. 2006), Jordan (41; Abdelqader et al. 2007) and Burkina Faso (34; Kondombo et al. 2003).

Table 2.7 Flock size and performance traits (Means* ±SD) of local chicken as given by 163 smallholder farmers across six major agro-ecological zones (AEZ) of Oman. See Materials and methods section 2.2.1 for AEZ name abbreviations.

AEZ Variable	MU n=30	BT n=30	NH n=30	EH n=28	EC n=25	DF n=20	Overall mean
Chicken per flock (n)	23.7 ±9.33	28.7 ±7.65	25.6 ±8.35	23.2 ±8.70	14.6 ±2.10	25.7 ±7.64	21.9 ±7.69
Age at first egg laying (weeks)	26.6 ±3.72	20.7 ±3.44	24.3 ±4.32	26.2 ±1.10	27.0 ±2.65	20.0 ±1.21	24.1±1.33
Clutch size (eggs)	11.4 ±3.33	13.2 ±3.33	13.1 ±2.05	13.0 ±2.00	10.1 ±1.16	13.2 ±1.97	12.3 ±2.85
Clutches per year (n)	5.2 ±0.12	5.0 ±0.41	4.8 ±0.22	6.0 ±0.48	5.3 ±0.27	5.1 ±0.45	5.2 ±0.23
Yearly egg production (n/hen)	59.3 ±7.24	66.0 ±10.05	62.9 ±6.66	78.0 ±6.61	53.5 ±7.21	67.3 ±5.71	64.5 ±6.91
Yearly hatchability (% eggs per hen)	86.5^{a} ±4.65	87.8^{ab} ±5.77	89.5^{ab} ±5.90	92.9^{b} ±7.16	85.4^{ab} ±5.64	86.5^{a}±6.11	88.1 ±6.01
Male : female ratio (m / 10 f)	2.4 ±0.86	2.4 ±1.08	2.2 ±1.01	2.3 ±1.02	1.9 ±0.93	2.0 ±0.90	2.1 ±0.92
Yearly mortality rate in flock (%)	15.6^{a} ±1.66	16.6^{abc} ±1.17	15.8^{a} ±1.76	16.2^{abc} ±1.96	16.9^{bc} ±0.73	17.0^{c}±1.21	16.3 ±1.37

* Within rows, means with different superscripts differ at $P<0.05$ between agro-ecological zones (Kruskal-Wallis test).

The results of the multiple linear regression analysis (Table 2.8) showed that family size, female gender, total livestock numbers (in Tropical Livestock Units (TLU); see footnote to Table 2.8), availability of a solid chicken house and the number of management assets used had a positive and significant influence on chicken flock size ($P<0.05$). The effect of family size on flock size might be explained by the importance of the chicken as an easy source of food for family needs. Mandleni and Anim (2012) stated that a larger family is more inclined to keep more livestock and chickens than a smaller family.

Gueye (2000) suggested that poultry, by its proximity to the homestead, is an obvious enterprise for women. The positive effect of female ownership on chicken flock size may be explained by the regular provision with leftovers of family meals which are mostly collected by women. The role of rural women in chicken husbandry and the important contribution of chickens to the livelihoods of rural households have been highlighted in several studies (Mapiye et al. 2008; Fisseha 2009). Thus, strategies for improving chicken productivity should consider women as the entry point and actively involve them in measures of improvement and conservation of traditional poultry breeds (Dessie and Ogle 2001).

Table 2.8 Coefficients of the multiple linear regressions predicting yearly chicken flock size, total egg production and yearly survival rates for local chickens of 163 smallholder farmers across six different agro-ecological zones of Oman.

Regression coefficients	b	SE_b	t-value[2]	Partial R^2	P≤
Dependent variable: Chicken flock size (n)					
Constant a (and SE_a)	4.57	2.08	−0.21	-	0.030
Family size (n)	0.38	0.13	2.89	0.15	0.004
Gender of chicken owner (female = 1, male = 0)	6.88	0.97	7.05	0.41	0.001
Total livestock (TLU1)	0.46	0.11	4.27	0.23	0.001
Using a solid house (1 = yes)	3.99	1.20	3.31	0.21	0.001
Management assets used (n)	0.80	0.42	1.90	0.12	0.059
Overall R^2	0.64				0.001
Dependent variable: Total egg production per hen (eggs/yr)					
Constant a (and SE_a)	13.37	4.01	3.34	-	0.001
Experience in chicken keeping (years)	2.90	0.23	12.88	0.70	0.001
Age of householder (1, >70 years)	3.27	1.68	1.95	0.11	0.053
Using a solid-stable house (1 = yes, 0 = no)	−6.10	2.22	−2.74	−0.17	0.007
Frequency of supplement feeding per day (n)	2.09	1.00	1.88	0.90	0.038
Chicken flock size (n)	0.22	0.11	2.09	0.11	0.050
Overall R^2	0.58				0.001
Dependent variable: Yearly survival rate of birds (%)					
Constant a (and SE_a)	81.91	0.32	254.4	-	0.001
Existence of hired laborers (1 = yes, 0 = no)	0.87	0.40	2.16	0.12	0.032
Age of householder (1, >70 years)	−0.75	0.39	−1.90	−0.10	0.059
Administration of medicine (1 = yes, 0 = no)	6.02	0.45	13.31	0.71	0.001
Overall R^2	0.60				0.001

[1]TLU: Tropical Livestock Unit, hypothetical animal of 250 kg live weight. Conversion factors used: cattle = 0.80, sheep and goats = 0.10, donkey =0.5, chicken=0.01.

t-value: A high absolute t-value suggests that a predictor variable is having a large impact on the dependent variable.

2.4 Conclusions and implications

Across Oman's different agro-ecological zones, rural chicken are exposed to insufficient feeding and housing, leading to a low productivity of laying hens. Since proper housing and cleaning, supplement feeding and health care substantially improve chicken performance, such measures must be promoted through training and extension programs. Given that chicken ownership, care and decision-making is largely in the hands of rural women, they have to be involved in development and conservation programs for local chicken in Oman. In view of the high variation in phenotypic and morphometric traits of regional chicken populations, any conservation program must be preceded by a comprehensive study of the genetic diversity of these populations so as to determine whether phenotypic dissimilarity is underpinned by genetic variation that can be deployed for such endeavors.

Acknowledgements

We are very grateful to the farmers who participated in our study, and to the members of Agricultural Directorates and Research Centers for providing farmer names and coordinating the field visits. Financial support for this study was provided by Sultan Qaboos University Muscat, Oman, through the Department of Animal and Veterinary Science (HM Fund SR/AGR/ANVS/08/01).

References

Abdelqader A., Wollny C.B. and Gauly M. (2007) Characterization of local chicken production systems and their potential under different levels of management practices in Jordan. *Tropical Animal Health and Production* **39,** 155-164.

Adebayo O. and Adeola R. (2005) Socio-economic factors affecting poultry farmers in Ejigbo local government area of Osun State. *Journal of Human Ecology* **18**, 39-41.

Aini I. (1990) Indigenous chicken production in South-East Asia. *World Poultry Sciences Journal* **46**, 51-57.

Al-Qamashoui B., Simianer H., Kadim I. and Weigend S. (2014a) Assessment of genetic diversity and conservation priority of Omani local chickens using microsatellite markers. *Tropical Animal Health and Production;* published online in March 2014; DOI: 10.1007/s11250-014-0558-9.

Al-Qamashoui B., Al-Ansari A., Simianer H., Weigend S., Mahgoub O., Costa V., Weigend A., Al-Araimi N. and Beja-Pereira A. (2014b) From India to Africa across Arabia: An mtDNA assessment of the origins and dispersal of chicken around the Indian Ocean Rim. *BMC Genetics* (*Submitted in Feb. 2014*).

Archer K.J. and Lemeshow S. (2006) Goodness-of-fit test for a logistic regression model fitted using survey sample data. *The Stata Journal* **6**, 97-105.

Badubi S.S., Rakereng M. and Marumo M. (2006) Morphological characteristics and feed resources available for indigenous chickens in Botswana. *Livestock Research Rural Development* **18**, 3. Accessed in Mar. 2012.

Barua A. and Yoshimur Y. (1997) Rural poultry keeping in Bangladesh. *Journal of World's Poultry Sciences* **53**, 387-394.

Biagi P. (2006) The shell-middens of the Arabian Sea and Persian Gulf: Maritime connections in the 7th Millennium BP. *Adumatu* **14**, 7-16.

Boivin N. and Fuller D.Q. (2009) Shell middens, ships and seeds: Exploring coastal subsistence, maritime trade and the dispersal of domesticates in and around the ancient Arabian Peninsula. *Journal of World Prehistory* **22**, 113-180.

Daikwo I., Okpe A. and Ocheja J. (2011) Phenotypic characterization of local chicken in Dekina. *International Journal of Poultry Sciences* **10**, 444-447.

Dessie T. and Ogle B. (2001) Village poultry production systems in the Central Highlands of Ethiopia. *Tropical Animal Health and Production* **33**, 521–537.

DGALR (Direcorate General of Agriculture and Livestock Research) (2011) Agriculture and livestock, five-year research strategy (2011-2015). MAF (Ed.), Muscat, Oman.

Egahi J., Dim N., Momoh O. and Gwaza D. (2010) Variations in qualitative traits in the Nigerian local chicken. *International Journal of Poultry Sciences* **10**, 978-979.

Fisseha M. (2009) Studies on production and marketing systems of local chicken ecotypes in Bure Woreda, north-west Amhara Regional State, Ethiopia. *MSc thesis, Hawassa University, Ethiopia.*

Gondwe T. (2004) Characterization of local chicken in low input-low output production systems: is there scope for appropriate production and breeding strategies in Malawi? *PhD thesis, Georg-August-Universität Göttingen, Germany.*

Gueye E. (2000) Women and family poultry production in Africa. *Development in Practices.* **10**, 98-102. Oxfam UK and Ireland.

Hair J.F., Black B., Babin B., Anderson R.E. and Tatham R.L. (2006) *Multivariate Data Analysis*, 6th ed. Upper Saddle River, NJ USA, Pearson Prentice Hall.

Hall H.T.B. (1986) *Diseases and parasites of livestock in the tropics.* 2nd ed. 328 p. Longman Group Ltd., Harlow, Essex, UK.

Jens R., Anders P., Charlotte V., Ainsh M. and Lone F. (2004) *Keeping village poultry; A technical manual for small-scale poultry production.* 89 p. Network for Smallholder Poultry Development, Copenhagen, Denmark.

Jugessur V., Pillay M., Ramnauth R. and Allas M. (2006) The socio-economic importance of family poultry production in the republic of Mauritius. In: *Improving farmyard poultry production in Africa: Interventions and their economic assessment*, pp. 164-178. IAEA-TECDOC-1489, Vienna,.

Kondombo S., Nianogo A., Kwakkel R., Udo H. and Slingerland M. (2003) Comparative analysis of village chicken production in two farming systems in Burkina Faso. *Tropical Animal Health and Production* **35**, 563-574.

Mandleni B. and Anim F. (2012) Climate change and adaptation of small-scale cattle and sheep farmers. *African Journal of Agriculture Research* **7**, 2639-2646.

Mapiye C., Mwale M., Mupangwa J., Chimonyo M., Foti R. and Mutenje M. (2008) A research review of village chicken production constraints and opportunities in Zimbabwe. *Asian-Australasian Journal of Animal Sciences* **11**, 1680-1688.

Mcainsh C., Kusina J., Madsen J. and Nyoni O. (2004) Traditional chicken production in Zimbabwe. *World's Poultry Sciences Journal* **60**, 233-246.

MoNE (Ministry of National Economy) (2010) Oman Census, data and indicators of the population. Muscat, Oman.

Mwalusanya N., Katule A. Mutayoba S., Mtambo M., Olsenand J. and Minga U. (2002) Productivity of local chicken under village management conditions. *Tropical Animal Health and Production* **34**, 405-416.

Petrus N., Mpofu I. and Lutaaya E. (2011) The care and management of indigenous chicken in northern communal areas of Namibia. *Livestock Research and Rural Development* **23**, 12. Accessed in Mar. 2012.

Ramlah A. (1996) Performance of village fowl in Malaysia. *World's Poultry Science Journal* **52**, 76-79.

Sainsbury D. (1992) Poultry health and management - chicken, turkeys, ducks, geese, quail. 3rd ed. 216 p., Blackwell Scientific Ltd, Oxford, UK.

Ssewannyana E., Ssali A., Kasadha T., Dhikusooka M., Kasoma P., Kalema J., Kwatotyo B. and Aziku L. (2008) On-farm characterization of indigenous chicken in Uganda. *Journal of Animal and Plant Sciences* **2**, 33-37.

Van Kampen M. (1974) Physical factors affecting energy expenditure. Morris T.R. and Freeman B.M. (Eds.), *Energy requirements of poultry*, pp. 47-59. Longman Group Ltd, Edinburgh, UK.

Yakubu A., Kuje D. and Okpeku M. (2009) Principal components as measures of size and shape in Nigerian indigenous chicken. *Thai Journal of Agricultural Sciences* **42**, 167-176.

Table S2.1 Color variation in body plumage, skin, beak, iris, shank and comb, and feather and comb type as determined in 199 local chicken across six agro-ecological zones (AEZ) of Oman. Values are numbers of birds per AEZ and sex showing the respective trait.

AEZ	MU		BT		NH		EH		EC		DF	
Sex (Female/Male)	F	M	F	M	F	M	F	M	F	M	F	M
Birds (n)	(30)	(6)	(30)	(6)	(30)	(6)	(28)	(6)	(25)	(6)	(20)	(6)
Phenotypic trait												
Neck color												
Black	7	0	8	0	9	0	6	1	6	0	2	0
White	3	1	5	0	2	1	6	0	4	1	1	0
Deep dark brown	8	2	0	2	4	1	0	0	3	1	8	2
Pale brown	6	0	15	0	12	1	6	0	4	0	1	2
Shining orange-yellow	6	3	2	4	3	3	10	5	8	4	8	2
Breast color												
Black	4	3	4	2	3	3	0	3	1	3	2	2
White	1	0	4	0	2	0	6	0	4	1	2	0
Deep dark brown	6	1	6	0	7	0	7	1	5	0	6	2
Pale brown	15	0	6	2	14	2	6	0	2	0	10	2
Shining orange-yellow	4	2	10	2	4	1	9	2	13	2	0	0
Wing color												
Black	3	1	4	1	2	0	0	0	1	1	0	0
White	1	0	2	1	2	0	6	0	1	0	0	1
Deep dark brown	0	1	4	0	8	1	6	1	6	0	6	1
Pale brown	13	1	4	0	6	1	6	2	3	1	8	1
Shining orange-yellow	3	1	6	2	4	2	9	1	8	3	0	1
Brown/black	10	2	10	2	8	2	1	2	6	1	6	2
Body feather type												
Normal firm	28	6	13	6	23	3	20	3	20	3	7	6
Many soft and fluffy	2	0	17	0	3	2	4	3	4	3	9	0
Few, skin showing	0	0	0	0	4	1	4	0	1	0	4	0
Skin color												
Yellow	0	2	1	2	1	2	0	3	0	1	3	1
Very light/pink	21	4	22	1	20	2	26	3	20	3	14	1
Very dark/black	9	0	7	3	9	2	2	0	5	2	3	4
Beak color												
Yellow	17	2	7	3	27	3	25	4	17	3	12	0
Beige to light brown	4	0	11	1	0	2	0	2	0	3	1	0
Black to very dark horn	9	4	12	2	3	1	3	0	8	0	7	6
Comb color												
Red	26	5	21	4	22	6	23	5	20	6	15	4
Black to very dark red/blue	4	1	9	2	8	0	5	1	5	0	5	2
Iris color												
White/yellow	0	0	2	0	0	1	0	0	0	1	1	0
Orange/red	20	4	20	2	22	3	27	4	19	4	13	3
Brown/black	10	2	8	4	8	2	1	2	6	1	6	3
Shank color												
Yellow	5	2	3	1	2	3	11	3	1	0	4	4
White	11	1	10	1	7	1	11	0	12	0	3	1
Blue-gray	14	3	14	4	16	1	6	0	6	2	10	0
Black	0	0	3	0	5	1	0	3	6	4	3	1
Comb type												
No comb	2	0	10	0	7	0	0	0	5	0	3	0
Single	25	4	17	6	19	2	25	3	20	6	15	3
Pea	3	2	0	0	0	4	3	3	0	0	0	3
V-shape	0	0	3	0	2	0	0	0	0	0	1	0
Butterfly	0	0	0	0	2	0	0	0	0	0	1	0

CHAPTER 3

Assessment of genetic diversity and conservation priority of Omani local chickens using microsatellite markers

B. Al-Qamashoui[1], H. Simianer[2], I. Kadim[3], and S. Weigend[4]

[1] *Animal Husbandry in the Tropics and Subtropics, University of Kassel and Georg-August-Universität Göttingen, Albrecht-Thaer-Weg 3, 37075 Göttingen, Germany*

[2] *Animal Breeding and Genetics, Georg-August-Universität, Göttingen, Albrecht-Thaer-Weg 3, 37075 Göttingen, Germany*

[3] *Department of Animal and Veterinarian Sciences, College of Agricultural and Marine Sciences, Sultan Qaboos University, Po Box 50, 123 Muscat, Sultanate of Oman*

[4] *Institute of Farm Animal Genetics, Friedrich- Loeffler-Institut, Höltystr. 10, 31535 Neustadt-Mariensee, Germany*

Published in a slightly modified version in:

Tropical Animal Health and Production

Published online in March 2014, DOI: 10.1007/s11250-014-0558-9

Abstract

Designing strategies for conservation and improvement of livestock should be based on assessment of genetic characteristics of populations under consideration. In Oman, conservation programs for local livestock breeds have been started. The current study assessed the genetic diversity and conservation potential of local chickens from Oman. Twenty-nine microsatellite markers were analysed in 158 birds from six agro-ecological zones: Batinah (BT), Dhofar (DF), North Hajar (NH), East Hajar (EH), Musandam (MU), and East Coast (EC). Overall, a total of 217 alleles were observed. Across populations, the average number of alleles per locus was 7.48 and ranged from 2 (MCW98 and MCW103) to 20 (LEI094). The mean expected heterozygosity (H_E) was 0.62. Average fixation index among populations (F_{ST}) was 0.034, indicating low population differentiation while the mean global deficit of heterozygotes across populations (F_{IT}) was 0.159. Based on Nei's genetic distance a neighbor-joining tree was constructed for the populations, which clearly identified the Dhofar population as the most distant one of the Omani chicken populations. The analysis of conservation priorities identified the Dhofar and Musandam populations as the ones that largely contribute to the maximal genetic diversity of the Omani chicken gene pool.

Keywords: Genetic diversity, conservation, microsatellites, Omani chicken.

3.1 Introduction

From all livestock species, chickens are most commonly distributed in rural and semi-urban regions in the tropics and subtropics; their husbandry is mostly characterized by free-range production systems (Mwalusanya et al. 2002; Jens et al. 2004). Birds under this production system satisfy most of their nutritional needs from scavenging. They are assumed to have developed adaptive features that contribute to survivability and reproduction of chickens under harsh climates, such as high temperatures, few management assets and poor diets (Gueye 2000; Msoffe et al. 2001; Hanotte and Jianlin 2005; Halima 2007).

The loss of genetic diversity within indigenous chickens and other farm animals has been a major concern as many local breeds are at risk of extinction (Gandini and Oldenbroek 1999). A significant number of indigenous chickens in the developing world is in danger of losing specific genetic features and variability, either by inbreeding or through crossbreeding with commercial breeds (Frankel and Soule 1981; Scherf 2000; FAO 2004). Consequently, this urges the implementation of conservation programs for local breeds and populations with the aim to reduce the increase in inbreeding and maintain a high level of genetic variation as a prerequisite to respond to future changes by genetic selection (Allendorf and Luikart 2007). In this regard, considering both within- and between-population genetic diversity assessed by molecular analyses is important for any conservation strategy (Caballero and Toro 2002; Simianer et al. 2003; Fernandez et al. 2008).

Assessing the genetic diversity using molecular markers within and among local breeds is an important step towards the estimation of populations' contribution to genetic diversity, which is a prerequisite for any conservation program (Boettcher et al. 2010). Among different molecular markers, microsatellites have been frequently used to assess and understand the genetic variation between populations, and their use in chicken populations is extensively documented (Crooijmans et al. 1996; Takahashi et al. 1998; Zhou and Lamont 1999; Zhang et al. 2002; Hillel et al. 2003; Granevitze et al. 2007; Berthouly et al. 2008).

In Oman, despite the important role that local chickens play in the smallholder families' economy (Saleh 2000; Kadim et al. 2009; MAF 2013; Al-Qamashoui et al. 2014), there have been a few studies on improvement of productivity (Saleh 2000; Kadim et al. 2009) and no studies on genetic diversity of these populations. This is even more dramatic as in the 1990s the Omani government, through the Ministry of Agriculture and Fisheries, started to import highly selected commercial chicken breeds, aiming to encourage the private sector to increase the productivity of chickens (DGLAR 2011). Since this importation and introduction of experimental lines, uncontrolled crossbreeding of local populations with these commercial chickens has occurred, accompanied by a serious threat to the gene pool of the local chicken in Oman (Kadim et al. 2009).

Oman is the third largest country on the Arabian Peninsula. It is located in the southeastern part of the peninsula between latitudes 16°40' and 26°20' north and longitudes 51°50' and 59°40' east with a coastline extending for 3165 km (DGMAN 2012). Its climate is characterized as hot and arid with a full-day average temperature of 38°C in the summer season. Average annual precipitation is 76.9 mm in the central region and 181.9 mm in the Dhofar mountains (south), and 117.4 mm/year across the whole country (DGMAN 2012). Its geographic location and the wide variation in climatic and landscape features are the reason for the country's abundant and unique faunal and floral biodiversity in the different agro-ecological zones (AEZ) (Al-Zidjali 1996; Al-Mashakhi and El-Hag 2007; Al-Saadi 2013).

Recently, Omani authorities have launched a national conservation initiative (DGLAR 2011; Mahgoub 2012; Al-Saadi 2013) aiming in the first step at identifying unique and valuable livestock genetic resources. The current study is part of this initiative and aims to (*ii*) evaluate the evolutionary relatedness of local Omani chicken populations with global reference chicken breeds (commercial and wild), and (*iii*) investigate the contribution of local populations to the total genetic diversity of Omani chickens for future conservation programs.

3.2 Materials and methods

3.2.1 Study area, sampling and DNA isolation

Six local chicken populations – Batinah (BT), North Hajar (NH), East Hajar (EH), Musandam (MU), East Coast (EC) and Dhofar (DF) – each representing a different agro-ecological zone (AEZ) of Oman, were selected for this study (Figure 2.1). A total of 158 female birds were collected from 158 local farms in 18 villages with an average of 19 - 30 samples from each AEZ. Villages close to big cities and commercial harbors were avoided. One milliliter of venous blood was collected from each individual into 1.5 mL tubes with EDTA as anticoagulant. The tubes were kept in a portable, insulated cool box with dry ice packs to maintain their temperature at 0-4° C until transport to a regional veterinary clinic. The samples were then transported in cool boxes to the lab work in Sultan Qaboos University where they were kept at -80°C until DNA isolation. Isolation of DNA was done using a standard phenol-chloroform method. DNA samples were genotyped at 29 microsatellite loci. Twenty-eight of these markers are part of the 30 microsatellites recommended for biodiversity studies in chickens (FAO 2004), while the two loci MCW0284 and LEI0192 were not used in the current study. Instead, locus MCW0080 was added to the set of markers. Multiplex PCR was carried out according to FAO recommendations (http://dad.fao.org/en/refer/library/guidelin/marker.pdf). Alleles of each locus were visualized as DNA fragments by 8% polyacrylamide gel electrophoresis using a LI-COR DNA analyzer. Allele-size scoring was performed with RFLPscan software package (Scanalytics, LI-COR), and standard alleles were loaded to each gel to adjust allele scoring across gel runs.

3.2.2 Statistical analyses

Microsatellite toolkit (Park 2001) was used to calculate total number of alleles, allele frequencies per population, average number of alleles per locus and observed (H_O) and expected (H_E) heterozygosities. Wright's F-statistics components (F_{IS}, F_{IT} and F_{ST}) were calculated using Genepop software (Raymond and Rousset 1995). Standard errors of the fixation indices were generated using jackknifing over loci based on 5000 iterations. F-statistics, H_E and H_O were calculated locus-wise using GenAlEx (Peakall and Smouse 2006). Based on pairwise Nei's genetic distances, an unrooted neighbor-joining tree (NJ) was constructed using Phylip software package (Felsenstein 1995) for the six local populations with eleven global reference populations selected from the AVIANDIV project (Weigend et al. 1998). These populations consisted of two commercial broiler dam lines (BRD_A and BRD_D), two commercial broiler sire lines (BRS_A and BRS_B), three commercial brown egg layer lines (BL_A, BL_C and BL_D), and two commercial white egg layer lines (WL_A and WL_C) as well as two wild populations, *Gallus gallus gallus* (RJFG) and *Gallus gallus spadiceus* (RJFSC). Each of the reference populations comprised 30 individuals. An additional NJ tree was constructed exclusively for the six Omani chicken populations. Statistical robustness of nodes for both NJ trees ***was assessed by 1000 bootstrap replicates***.

In order to rank the populations and assess their relative importance for conservation of genetic diversity, we used two methods described by Caballero and Toro (2002) as implemented in the software package Metapop (Perez-Figueroa et al. 2009). Before running the analyses, we grouped the populations into 4 groups according to their clustering in NJ tree and geographical adjacency; NH+BT, EC+EH, MU and DF (Figure 3.1 b). The first method evaluates the loss (or gain) of total genetic diversity (GD_t) when it is recalculated excluding a single population. So, for each population i, the global genetic diversity GD_{tli} is calculated excluding this population, then the loss of genetic diversity (L_{GD}) is calculated as

$$L_{GD} = \left(\left(GD_t - GD_{tli}\right) / GD_t\right) \times 100$$

The higher the contribution of a population to the global genetic diversity is, the higher will be the decrease of genetic diversity when that population is excluded from the analysis. Negative values indicate that global genetic diversity is increased when that population is excluded. The same procedure was performed for both the within and between population components of genetic diversity.

The second method searches for the relative contribution (c_i) that a population i will add to a pool (like a synthetic population) with maximum genetic diversity $(GDpool)$ by using the Simulated Annealing searching algorithm (Kirkpatrick *et al.* 1983), This is achieved by obtaining the values of c_i that maximize $GDpool$ as

$$GDpool = 1 - \sum_{i=1}^{n} c_i \left[f_{ii} - \sum_{j=1}^{n} D_{ij} c_j \right]$$

where c_i is restricted to $c_i \geq 0$ and $\sum_{i=1}^{n} c_i = 1$ with n being the number of populations, f_{ii} the coancestry coefficient of population i, and D_{ij} Nei's minimum distance between population i and j (Caballero and Toro 2002).

3.3 Results

3.3.1 Microsatellite markers, population diversity and relationship

The 29 analyzed microsatellite loci rendered a total number of 217 alleles (Table S3.1). All loci were found to be polymorphic. The number of alleles per locus ranged from 2 (MCW98 and MCW103) to 20 (LEI094) with a global mean of 7.48. The mean H_O and H_E across loci were 0.54 and 0.61, respectively.

Table 3.1 Summary statistics (± SD) computed per population for 29 microsatellite loci; Mean number of alleles (MNA), mean expected (H_E) and observed (H_O) heterozygosity and F_{IS}. See Material and methods section 3.2.1 for population name abbreviations.

Population	No. of samples	MNA	H_E	H_O	F_{IS}*
BT	30	5.1 ± 2.23	0.57 ± 0.04	0.53 ± 0.02	0.072
DF	19	5.0 ± 1.86	0.66 ± 0.02	0.53 ± 0.02	0.193
EC	24	5.2 ± 2.37	0.60 ± 0.04	0.51 ± 0.02	0.152
EH	28	5.5 ± 2.50	0.62 ± 0.03	0.53 ± 0.02	0.152
MU	27	5.4 ± 2.43	0.65 ± 0.03	0.53 ± 0.02	0.181
NH	30	5.5 ± 2.91	0.61 ± 0.03	0.58 ± 0.02	0.056
Overall mean		5.3 ± 2.43	0.62 ± 0.03	0.54 ± 0.02	0.134

* The F_{IS} values in all populations were significantly different from zero ($p<0.05$).

The average number of alleles per locus and population was 5.3 ± 2.43 and ranged from 5.0 (DF) to 5.5 (EH and NH) (Table 3.1). Mean H_E across all populations was 0.62 ± 0.03 with the lowest value obtained from BT (0.57 ± 0.04) and the highest from DF (0.66 ± 0.02). Average F_{IS} was 0.134 ranging from 0.056 (NH) to 0.193 (DF). In all populations, F_{IS} values were significantly different from zero ($p<0.05$). Pairwise F_{ST} between the Omani local chicken subpopulations varied between 0.011 (BT and NH) and 0.066 (BT and DF) with an average value of 0.034 (Table 3.2).

Table 3.2 Estimated pairwise F_{ST} as a measure of genetic differentiation among six local Omani chicken populations. See Materials and methods section 3.2.1 for population name abbreviations.

POP	BT	DF	EC	EH	MU	NH
BT						
DF	0.066					
EC	0.019	0.053				
EH	0.028	0.045	0.012			
MU	0.032	0.035	0.021	0.015		
NH	0.011	0.052	0.017	0.020	0.019	

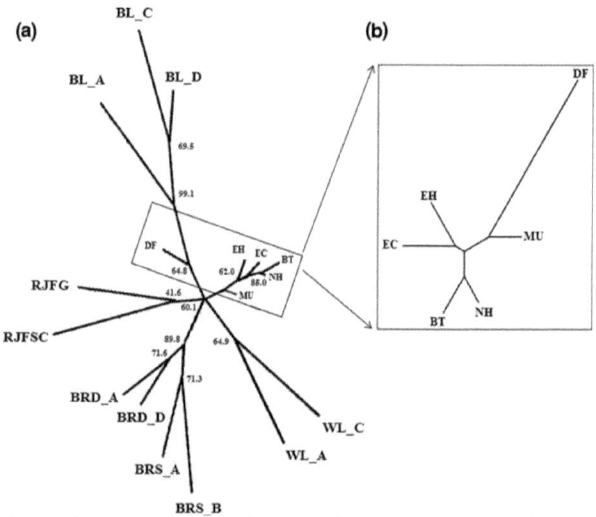

Figure 3.1 Neighbor-joining tree for (a) 17 chicken populations (6 Omani and 11 reference populations) and (b) the six Omani chicken populations, based on Nei's standard genetic distance. Numbers in nodes are percentage bootstrap values obtained from 1,000 replicates. See Material and methods section 3.2.1 for population name abbreviations.

Including reference populations (Figure 3.1a), a neighbor-joining tree analysis based on Nei's standard genetic distance revealed five groups: one cluster consisting of five populations from the northern part of Oman (BT, NH, EH, EC and MU) with a separated branch of DF from southern Oman, and four additional clusters of brown egg layers (BL_A, BL_C and BL_D), wild breeds (RJFG and RJFSC), broilers (BRD_A, BRD_D, BRS_A and BRS_B) and white egg layers (WL_A and WL_C). When focusing on the Omani populations only (Figure 3.1b), NH and BT clustered in one branch, while EC and EH were in another branch. DF and MU formed a third branch, but DF divided from MU due to its high genetic distance.

3.3.2 Relative importance of populations for conservation

Results of assessing the impact of removing one population from the Omani local chicken gene pool and calculating the proportional contribution of each population (in %) to a pool with maximal genetic diversity (GD_{pool}) are shown in Table 3.3 (a) and (b). When genetic diversity within populations was considered, removal of BT+NH group increased the genetic diversity by 2.04%, while removal of MU decreased the diversity by 0.82%. When genetic diversity between populations was considered, the highest impact on the diversity was observed when DF was removed. Taking into consideration both components, the total gene diversity showed the highest loss when DF was removed (2.28%), whereas removal of BT+NH, followed by EC+EH, resulted in a gain of diversity by 1.93% and 0.54%, respectively (Table 3.3a). Table 3.3b displays the proportional contributions of populations to a synthetic gene pool that maximizes GD_{pool} to 0.66% (+ 6.1). It shows that such a synthetic population of maximum diversity should be built with 52.9% of individuals from DF, 39.6% from MU, 4.5% from BT+NH and 3.0% from EC+EH.

Table 3.3 (a) Loss (+) or gain (-) of diversity components after removal of each population (in %) and (b) proportional contribution of each population (in %) to a pool with maximal genetic diversity (GD_{pool}). See Material and methods section 3.2.1 for population name abbreviations.

(a) Population removed	Within populations diversity	Between populations diversity	Total gene diversity	(b) Contribution to maximum GD_{pool}
BT+ NH	-2.04	+0.11	-1.93	4.5
EC+EH	+0.05	-0.59	-0.54	3.0
DF	+0.68	+1.60	+2.28	52.9
MU	+0.82	+0.13	+0.95	39.6
				GD_{pool} = 0.66

3.4 Discussion

Genetic characterization of livestock populations is a prerequisite in the decision making process for conservation measures (Boettcher et al. 2010). The current study used a set of 29 microsatellite markers which has also been used to evaluate genetic diversity of chicken populations in several previous studies (e.g., Granevitze et al. 2007; Muchadeyi et al. 2007; Chen et al. 2008; Bodzsar et al. 2009; Granevitze et al. 2009; Cuc et al. 2010; Mtileni et al. 2011). This allows direct comparison of the results obtained in this study to results reported previously.

Genetic diversity levels reported here for all loci and populations were comparable with findings reported by Muchadeyi et al. (2007) who found high degree of heterozygosity, and high number of alleles in local chicken populations of Zimbabwe and South Africa, respectively. Similar levels of heterozygosity were also reported for village chickens in West African countries based on a set of 22 microsatellite loci (Leroy et al. 2012). On the other hand, these figures were higher than those reported for pure breed lines (Muchadeyi et al. 2007) and for the Hungarian and European breeds (Bodzsar et al. 2009). Generally, low positive values of F_{IS} especially in NH and BT populations indicate a low heterozygote deficiency, which suggests little inbreeding within these populations. The average F_{ST} value of Omani chickens (0.034) was similar to that stated by Muchadeyi et al. (2007) for local African breeds. However, higher degrees of genetic differentiation (0.052-0.066) were observed between DF-NH, DF-EC and DF-BT. Accordingly, the NJ tree also showed a distant clustering of DF from the other Omani populations. The NJ tree showed a clear clustering of the remaining five Omani local chicken populations to a joint branch in the centre of the tree, suggesting little or no crossbreeding between the five studied Omani populations and commercial layer and broiler populations represented by the reference populations (Figure 3.1a). When focusing on Omani local populations, the NJ tree clustering clearly reflected the geographical neighboring between AEZ especially for NH and BT (Figure 3.1b and Figure 2.1), while the clustering of EC and EH could be attributed to the tribalism relationship and trade connectivity between farmers in both AEZ. Looking at the geographic distribution of the Omani breeds, MU are the most north, and DF the most south (Figure 2.1). Therefore, their genetic distinctiveness from other Omani chicken (Figure 3.1) may be attributed to their splitting from the more homogenous gene pool in the middle of Oman.

Our results demonstrated that Dhofar (DF) and Musandam (MU) populations showed largest contributions to between-population diversity (Table 3.3a). In addition, both populations showed reasonable contribution to the within-population diversity component. Therefore, DF and MU were ranked highest among other populations in the context of conservation priorities, aiming at a gene pool with maximum diversity in a synthetic population (Table 3.3b). This indicates the uniqueness and diversity features of these two populations and supports their importance for Omani chicken population conservation programs.

3.5 Conclusions

This study explored the genetic variation within Omani local chicken at the autosomal level and demonstrated an absence of substructuring across the agro-ecological zones. The findings of this study could be implemented by policy makers for initiating a conservation program. Overall, the gene pool of Omani chicken represents a rich legacy to the country's genetic resources. However, additional studies are needed to explore signatures of molecular adaptation of chicken to the extremely harsh environments in Oman.

Acknowledgements

We are grateful to the farmers who participated in our study, and thank the Agricultural Directorates and Research Centers in Oman for coordinating field visits. We highly appreciate the technical assistance of Mrs. Annett Weigend, Mr. Maik Przyklenk and Mrs. Natalie Janus at the Institute of Farm Animal Genetics, Mariensee. This research was embedded in a project aiming at the genetic characterization of Omani livestock resources, funded by the Sultan Qaboos University through HM Fund SR/AGR/ANVS/08/01.

References

Al-Mashakhi M. and El-Hag B. (2007) Oman, country pasture/forage resource profiles. FAO (Ed), p. 23, Rome, Italy.

Al-Qamashoui B., Mahgoub O., Kadim I. and Schlecht E. (2014) Towards conservation of Omani local chicken: phenotypic characteristics, management practices and performance traits. *Asian-Australasian Journal of Animal Sciences* **27** (6), 767-777.

Al-Saadi N. (2013) The Animal and Plant Genetic Resources Center; summary of the proposed strategy. In:. *Proceedings of the Symposium on biotechnology and conservation of species from arid regions*. SQU (Ed), Muscat, Oman, 10-13/Feb/2013.

Al-Zidjali T.M. (1996) Oman, country report to the FAO international technical conference on plant genetic resources. In: *FAO International Technical Conference on Plant Genetic Resources*. FAO (Ed), Leipzig, Germany, 17-23/Jun/1996.

Allendorf F.W. and Luikart G. (2007) *Conservation and the genetics of populations*. Blackwell, Oxford, UK.

Berthouly C., Bed'Hom B., Tixier-Boichard M., Chen C.F., Lee Y.P., Laloë D., Legros H., Verrier E. and Rognon X. (2008) Using molecular markers and multivariate methods to study the genetic diversity of local European and Asian chicken breeds. *Animal Genetics* **39**, 121-129.

Bodzsar N., Eding H., Revay T., Hidas A. and Weigend S. (2009) Genetic diversity of Hungarian indigenous chicken breeds based on microsatellite markers. *Animal Genetics* **4**, 516–523.

Boettcher P.J., Ajmone-Marsan P. and Lenstra J.A. (2010) Recent development in the genetic characterization of animal genetic resources. FAO (Ed.), Rome, Italy.

Caballero A. and Toro M.A. (2002) Analysis of genetic diversity for the management of conserved subdivided populations. *Conservation Genetics* **3**, 289-299.

Chen G.H., Bao W.B., Shu J., Ji C.L., Wang M.Q., Eding H., Muchadeyi F. and Weigend S. (2008) Assessment of population structure and genetic diversity of 15 Chinese indigenous chicken breeds using microsatellite markers. *Asian-Australasian Journal of Animal Sciences* **21**, 331-339.

Crooijmans R.P., Groen A.F., van Kampen A.J.A., van der Beek S., van der Poel J.J. and Groenen M.A.M. (1996) Microsatellite polymorphism in commercial broiler and layer lines estimated using pooled blood samples. *Poultry Science* **75**, 904-909.

Cuc N.T., Simianer H., Eding H., Tieu H.V., Cuong V.C., Wollny C.B., Groeneveld L.F. and Weigend S. (2010) Assessing genetic diversity of Vietnamese local chicken breeds using microsatellites. *Animal Genetics* **41**, 545-547.

DGALR (Direcorate General of Agriculture and Livestock Research) (2011) Agriculture and livestock, five-year research strategy (2011-2015). MAF (Ed.), Muscat, Oman.

DGMAN (Directorate General of Meterology and Navigation) (2012) Annual report of weather forecast for Oman. Muscat, Oman.

FAO (2004) The state of the world's animal genetic resources for food and agriculture. Rischkowsky B. and Pilling D. (Eds), Rome, Italy.

Felsenstein J. (1995) PHYLIP (Phylogeny Inference Package). Department of Genetics, University of Washington, Seattle, USA.

Fernandez J., Toro M.A. and Caballero A. (2008) Management of subdivided populations in conservation programs: development of a novel dynamic system. *Genetics* **179**, 683–692.

Frankel O.H. and Soule M.E. (1981) *Conservation and evolution*. Cambridge University Press, Cambridge, UK.

Gandini G.C. and Oldenbroek J.K. (1999) Choosing the conservation strategy. In: *Genebanks and the conservation of farm animal genetic resources*. Oldenbroek J.K. (Ed), pp. 11–31. DLO, Lelystad, Netherlands.

Granevitze Z., Hillel J., Chen G.H., Cuc N.T., Feldman M., Eding H. and Weigend S. (2007) Genetic diversity within chicken populations from different continents and management histories. *Animal Genetics* **38**, 576-583.

Granevitze Z., Hillel J., Feldman M., Six A., Eding H. and Weigend S. (2009) Genetic structure of a wide-spectrum chicken gene pool. *Animal Genetics* **40**, 686-693.

Gueye E. (2000) Women and family poultry production in Africa. *Development in Practice* **10**, 98-102.

Halima H.M. (2007) Phenotypic and genetic characterization of indigenous chicken populations in Northwest Ethiopia. *PhD Thesis, Department of Animal, Wildlife and Grassland Sciences, University of the Free State, Bloemfontein, South Africa.*

Hanotte O. and Jianlin H. (2005) Genetic characterization of livestock populations and its use in conservation decision-making. In: *International workshop: the role of biotechnology for the characterization and conservation of crop, forestry, animal and fishery genetic resources.* FAO (Ed.), Turin, Italy.

Hillel J., Groenen A.M., Biochard M.T., Korol A.B., David L., Kirzhner V.M., Burke T., Dirie A.B., Crooijmans R.P., Elo K., Feldman M.W., Freidlin P.J., Tanila A.M., Oortwijn M., Thomson P., Vignal A., Wimmers K. and Weigend S. (2003) Biodiversity of 52 chicken populations assessed by microsatellite typing of DNA pools. *Genetics Selection Evolution* **35**, 533-357.

Jens R., Anders P., Charlotte V., Ainsh M. and Lone F. (2004) *Keeping village poultry; A technical manual for small-scale poultry production.* 89 p. Network for Smallholder Poultry Development, Copenhagen, Denmark.

Kadim I., Al-Marzooqi W., Al-Waheebi S. and Mahgoub O. (2009) Improving egg and meat production of Omani local chicken through genetic and phenotypic selection programmes. Sultan Qaboos University, Muscat, Oman.

Kirkpatrick S., Gelatt C. and Vecchi M.P. (1983) Optimization by simulated annealing. *Science* **220**, 671–680.

Leroy G., Kayang B.B., Youssao I.A., Yapi-Gnaore C.V., Osei-Amponsah R., Loukou N.E., Fotsa J.C., Benabdeljelil K., Bed'hom B., Tixier-Boichard M. and Rognon X. (2012) Gene diversity, agroecological structure and introgression patterns among village chicken populations across North, West and Central Africa. *BMC Genetics* **13**, 34.

MAF (Ministry of Agriculture and Fisheries) (2013) Directorate of Rural Women; projects and programs of agricultural production and extension. Muscat, Oman.

Mahgoub O. (2012) Characterization, evaluation and conservation of indigenous animal genetic resources in the Sultanate of Oman, Research Prposal. Sultan Qaboos University, Muscat, Oman.

Msoffe P.L., Minga U.M., Olsen J.E., Yongolo M.G., Juul-Madsen H.R., Gwakisa P.S. and Mtambo M.M. (2001) Phenotypes including immunocompetence in scavenging local chicken ecotypes of Tanzania. *Tropical Animal Health and Production* **33**, 341-354.

Mtileni B.J., Muchadeyi F.C., Maiwashe A., Groeneveld E., Dzama K. and Weigend S. (2011) Genetic diversity and conservation of South African indigenous chicken populations. *Journal of Animal Breeding and Genetics* **128**, 209-218.

Muchadeyi F.C., Eding H., Wollny C.B., Groeneveld E., Makuza S.M., Shamseldin R., Simianer H. and Weigend S. (2007) Absence of population substructuring in Zimbabwe chicken ecotypes inferred using microsatellite analysis. *Animal Genetics* **38**, 332-339.

Mwalusanya N., Katule A., Mutayoba S., Mtambo M., Olsenand J. and Minga U. (2002) Productivity of local chicken under village management conditions. *Tropical Animal Health and Production* **34**, 405–416.

Park S.D. (2001) Trypanotolerance in West African cattle and the population genetic effects of selection. *PhD Thesis, Trinity College, University of Dublin, Ireland.*

Peakall R. and Smouse P.E. (2006) GENALEX 6: Genetic analysis in Excel; population genetic software for teaching and research. *Molecular Ecology Notes* **6**, 288-295.

Perez-Figueroa A., Saura M., Fernandez J., Toro M.A. and Caballero A. (2009) METAPOP - a software for the management and analysis of subdivided populations in conservation programs. *Conservation Genetics* **10**, 1097–1099.

Raymond M. and Rousset F. (1995) GENEPOP (version 1.2): population genetics software for exact tests and ecumenicism. *Journal of Heredity* **86**, 248-249.

Saleh F. (2000) Improving the local Dhofari poultry, Progress Report. Ministry of Agriculture and Fisheries, Dhofar, Sultanate of Oman.

Scherf B.D. (2000) *World watch list for domestic animal diversity*. FAO (Ed.) Rome, Italy.

Simianer H., Marti S.B., Gibson J., Hanotte O. and Rege J.E.O. (2003) An approach to the optimal allocation of conservation funds to minimize loss of genetic diversity between livestock breeds. *Ecological Economics* **45**, 377-392.

Takahashi H., Nirasawa K., Nagamine Y., Tsudzuki M. and Yamamoto Y. (1998) Genetic relationships among Japanese native breeds of chicken based on microsatellite DNA polymorphisms. *Journal of Heredity* **89**, 543–546.

Weigend S., Groenen M.A., Tixier-Boichard M., Vignal A., Hillel J., Wimmers K., Burke T. and Maki-Tanila A. (1998) Development of strategy and application of molecular tools to assess biodiversity in chicken genetic resources (AVIANDIV) EC Contract No. BIO4-CT98-0342.

Zhang X., Leung F., Chan D., Chen Y. and Wu C. (2002) Comparative analysis of allozyme, random amplified polymorphic DNA, and microsatellite polymorphism on Chinese native chickens. *Poultry Science* **81**, 1093-1098.

Zhou H. and Lamont S.J. (1999) Genetic characterization of biodiversity in highly inbred chicken lines by microsatellite markers. *Animal Genetics* **30**, 256-264.

Supplementary material

Table S3.1 Loci names, number of alleles, observed (H_O) and expected (H_E) heterozygosity calculated for each microsatellite across 6 Omani local chicken populations.

Locus	Allele range	No. alleles	H_O	HE
ADL112	122–134	4	0.41	0.46
ADL268	104–116	7	0.77	0.75
ADL278	114–123	7	0.56	0.72
LEI94	245–289	20	0.71	0.84
LEI234	216–368	16	0.7	0.85
LEI166	350–366	3	0.5	0.51
MCW330	256–290	6	0.62	0.67
MCW295	88–108	10	0.38	0.53
MCW248	207–223	3	0.33	0.4
MCW222	220–226	4	0.37	0.6
MCW216	137–149	6	0.33	0.46
MCW206	221–249	8	0.69	0.69
MCW183	296–326	10	0.67	0.67
MCW165	114–118	3	0.6	0.64
MCW123	76–94	10	0.48	0.52
MCW111	98–114	7	0.73	0.71
MCW104	190–228	11	0.44	0.5
MCW103	262–274	2	0.35	0.33
MCW98	261-265	2	0.28	0.32
MCW81	112–145	9	0.58	0.61
MCW80	266–282	12	0.45	0.7
MCW78	135–145	6	0.39	0.58
MCW69	158–176	8	0.75	0.73
MCW67	176–190	4	0.54	0.63
MCW34	214–246	12	0.66	0.75
MCW20	179–185	4	0.68	0.7
MCW16	170–204	8	0.7	0.72
MCW14	160–182	9	0.4	0.43
MCW37	154–160	6	0.47	0.59
Average		7.48	0.54	0.61

CHAPTER 4

From India to Africa across Arabia: An mtDNA assessment of the origins and dispersal of chicken around the Indian Ocean Rim

B. Al-Qamashoui[1], A. Al-Ansari[2], H. Simianer[3], S. Weigend[4], O. Mahgoub[5], V. Costa[6], A. Weigend[4], N. Al-Araimi[6], A. Beja-Pereira[6]

[1] *Animal Husbandry in the Tropics and Subtropics, University of Kassel and Georg-August-Universität Göttingen, Albrecht-Thaer-Weg 3, 37075 Göttingen, Germany*

[2] *Department of Biology, Sultan Qaboos University, Po.Box 50, 123 Muscat, Sultanate of Oman*

[3] *Animal Breeding and Genetics, Georg-August-Universität Göttingen, Albrecht-Thaer-Weg 3, 37075 Göttingen, Germany*

[4] *Institute of Farm Animal Genetics, Friedrich- Loeffler-Institut, Höltystr. 10, 31535, Neustadt-Mariensee, Germany*

[5] *Department of Animal and Veterinarian Sciences, Sultan Qaboos University, Po.Box 50, 123 Muscat, Sultanate of Oman*

[6] *Centro de Investigação em Biodiversidade e Recursos Genéticos (CIBIO), InBIO, Universidade do Porto, R. Padre Armando Quintas 7, 4485-661 Vairão, Portugal*

In a slightly modified form this paper has been submitted for publication to:

BMC Genetics

in February 2014

Summary

The Arabian Peninsula is thought to have played a major role in the diffusion of livestock across the Indian Ocean. However, very limited genetic data is available on the local chicken populations from this region. In this study, sequencing data from a fragment of the control region from the mitochondrial genome (mtDNA) from 175 individuals and 32 published sequences was used to assess genetic diversity and inference on the maternal origins of local chickens from the Arabian Peninsula (including the isle of Socotra) and Horn of Africa. Because of its role in the human movements between Asia and Africa, and to investigate the dispersal of chicken around the Indian Ocean Rim, sequences from Africa and India were also included in this study. We found a total of 27 haplotypes with an average haplotype diversity of 0.7588 (±0.0300), clustering into three of the previously identified phylogenetic clades. The most frequent observed haplotypes from the Arabian Peninsula (and Socotra) clustered in clade E, which is supposed to have originated on the Indian subcontinent. While samples from Somalia belong mostly to clade C, which supposedly has its roots in Southeast Asia, a few individuals, mostly from North Oman, clustered in clade A, originating from Southeast and/or East Asia. The wide presence of clade E on the Arabian Peninsula points towards a major influence of the Indus Valley center of origin in the genesis of Arabian local chicken. Isolation by distance tests showed that chicken diffusion across the Indian Ocean is correlated with the proximity to the two main centers of chicken domestication. The high frequency of haplotypes that originated from the Indian Subcontinent domestication event, in Arabia, provides interesting clues on the role of this Peninsula in the diffusion of livestock around the Indian Ocean rim.

Keywords: Arabian Peninsula; dispersal routes; Indian Ocean; local chicken; mtDNA.

4.1 Introduction

Indigenous chicken populations play various and crucial roles in the economy, socio-culture and sustaining of livelihoods across the world (Kryger et al. 2010). Many archaeological findings have shown that chicken have had a fast dispersion throughout the world (Williamson 2000; Storey et al. 2012; Wragg et al. 2012). Interestingly, several genetic studies using mitochondrial DNA (mtDNA) to access the dispersal pattern of chickens from their centers of origin, have shed light on prehistoric human migration, trade routes, and cross cultural diffusion (Gongora et al. 2008; Muchadeyi et al. 2008; Razafindraibe et al. 2008; Dana et al. 2010; Storey et al. 2012), showing that this livestock species is a very promising bioproxy to study the past contacts between different civilizations.

Despite the large archaeological and evolutionary evidences for a South-Asian origin of this species, the last two decades have been fertile in pinpointing the geographic origins of the domestic chicken (Fumihito et al. 1996; Liu et al. 2006; Kanginakudru et al. 2008; Storey et al. 2012; Miao et al. 2013). From all these studies, the one by Liu and colleagues (2006) was the first to demonstrate that several sub-species of red jungle fowl from South and Southeast Asia and surrounding areas were involved in the genetic makeup of modern chickens. Their mtDNA-sequencing data analysis from several hundred domestic chicken and jungle fowl species have demonstrated that all the mitochondrial diversity falls within nine different phylogenetic clades, which they have designated from A to I. The higher genetic variation and phylogenetic similarity between the domestic chicken and the jungle fowl observed in the Indian subcontinent pointed to this region as the center of origin for clades E, C and D. Southeast Asia, for the same reason, is the center of origin for clade A, and Southwest China origin of clade B (Liu et al. 2006; Kanginakudru et al. 2008; Miao et al. 2013). Since then, the scientific community adopted Liu's clade nomenclature to classify domestic chicken according to its origins.

The Indian Ocean is well known for its many maritime trading routes between the Indian subcontinent, Middle East and Africa since 2000-3000 B.C. (Edens 1992; Fuller et al. 2011). Frequently, seaborne trading has been considered responsible for the introduction and dispersal of chickens along East Africa, and indeed, clade E (following Liu's nomenclature) is the most frequent one observed in local chickens from this region (see review by Mwacharo et al. 2013).

The Arabian Peninsula is located at the northwest shore of the Indian Ocean at the cross road between the three major domestication centers (Near East, Africa and South Asia). Across times, this region has played an important role in human migration and in the dispersal of commodities, plants, crops and animals between India and Africa (Pickering 2007; Boivin et al. 2010; Groucutt and Petraglia 2012). Despite many archaeological (AbdulNayeem 2000; Potts 2000), historical and linguistic (Boivin et al. 2010), and genetic findings (Al-Abri et al. 2012; Badro et al. 2013; Mahgoub et al. 2013), that pointed to the paramount role of this Peninsula in the dispersion of farming species, very little is known about the genetic make-up of the local Arabian Peninsula livestock species.

Due to the growing concern regarding conservation and sustainable utilization of the genetic resources worldwide, the evaluation of animal genetic diversity and understanding their domestication and distribution history became an important issue. Because the control region of mitochondrial DNA (D-loop region of mtDNA) has a higher evolutionary rate when compared to genomic DNA, and sequence data are relatively easy to obtain, many local chicken populations have been assessed for this marker. In fact, a large plethora of these studies have contributed to the current understanding on the geographic distribution and origin of the domestic chicken across various regions of the world (Liu et al. 2006; Muchadeyi et al. 2008; Galtier et al. 2009; Silva et al. 2009; Mwacharo et al. 2011; Storey et al. 2012; Miao et al. 2013).

Despite being of paramount importance for smallholder family's economy on the Arabian Peninsula (Al-Yousef 2007; Kadim et al. 2009; Al-Qamashoui et al. 2014), nothing is known about the genetic origins and diversity of domestic chicken from this region. Here we aimed to (1) assess the maternal genetic origins of the native chicken populations from the entire Arabian Peninsula, and (2) compare the genetic origins of chicken from this region with others from around the Indian Ocean rim to reconstruct the main paths of chicken dispersion across this region.

4.2 Materials and methods

One milliliter of venous blood or a tiny piece of comb was collected from 175 village chickens from 11 sites representing 11 populations in Oman, Saudi Arabia, Yemen, Socotra Island and Somalia (as the nearest region to the Arabian Peninsula at the Horn of Africa). Blood samples were collected from the wing vein of Omani chickens into tubes with heparin or EDTA as anticoagulant and kept at -80° C until DNA isolation. Tiny tissue samples were collected from local chicken in Saudi Arabia, Yemen and Somalia and kept in tubes containing 96% ethanol until DNA extraction. Individuals were selected from village households and farms located far away from major cities and harbors. The details on the regions and sampling sizes are presented in Figure 4.1 and Table S4.1.

DNA was extracted using standard silica-column based commercial kits (D-Neasy Blood & Tissue Kit, Qiagen, UK). A 550 bp fragment from mtDNAD-loop region was amplified by PCR using two primers L16750 (5'-AGGACTACGGCTTGAAAAGC-3) and H522 (5'-ATGTGCCTGACCGAGGAACCAG-3'). PCR were performed in a 25 µl volume [1x reaction buffer, 75mM MgCl2, 5mM of each dNTP, 10pM of each primer, and 1U of Taq polymerase (SABC Inc.)] following 35 cycles of 1min at 94 °C, 1min at 63 °C, and 1min at 72 °C. PCR products were then purified and sequenced in both directions (forward and reverse) using the Big DyeTM Terminator v.3.1 Cycle Sequencing Ready Reaction on an ABI PRISM 3100 sequencer (Applied Biosystems, Warrington, UK). The generated raw sequences were edited and aligned with additional sequences using software package DNASTAR v.7.1 (DNASTAR Inc., Madison, WI, USA) and aligned using MEGA software (Tamura et al. 2011). Extra nucleotide bases were trimmed from all sequences to make a homogeneous length of 420 bp. In addition, 32 sequences from chicken from Saudi Arabia (KC436009 - KC436040) were retrieved from GenBank (http://www.ncbi.nlm.nih.gov/genbank) and aligned with those produced by us.

The studied populations were grouped in five different groups according to their geographical location: (1) Northeast Arabia (NEA); (2) Southeast Arabia (SEA); (3) Central Arabia (CTA); (4) Socotra Island (SOC) and (5) Horn of Africa (HAF). The details of these groups and sampling areas are given in Figure 4.1.

ARLEQUIN software v.3.5 (Excoffier and Lischer 2010) was used to calculate the intra-group diversity measures such as number of segregating sites, number of haplotypes, nucleotide and haplotype diversities. Nucleotide diversity (π) is defined as the average number of nucleotide differences per site between any two DNA sequences chosen randomly from the same group (Nei 1978) and measured as

$$\pi = \sum_{ij} x_i x_j \pi_{ij}$$

where x_i and x_j are the respective frequencies of the i_{th} and j_{th} sequences; π_{ij} is the number of nucleotide differences per nucleotide site between the i_{th} and j_{th} sequences. Haplotype diversity

(H) is defined as the probability that two haplotypes drawn uniformly at random from the group are not the same (Nei and Tajima 1981) and is measured as

$$H = \frac{n}{n-1}\left(1 - \sum_i x_i^2\right)$$

where x_i is the haplotype frequency of each haplotype in the sample; and n is the sample size. Genetic differentiation among populations was calculated using population pairwise (Φ_{ST}) values implemented in ARLEQUIN v.3.5 and tested for significance by permutating the haplotypes or individuals between the populations using 10,000 permutations at the 0.05 significance level. To investigate the evolutionary relationships between chicken haplotypes of the Arabian Peninsula and the Horn of Africa we have traced the most frequent haplotypes to their population source (domestication center) as proposed by Liu et al. (2006). Several Median-Joining Networks were constructed following the algorithms of Bandelt et al. (1995) using the program NETWORK v.4.1 (http:// www.fluxus-engineering.com/sharenet.htm). For network analysis, epsilon parameter was set to 0 and all characters were given the same weigh (10).

Frequencies of clade E in chicken populations around the Indian Ocean rim were retrieved from recent studies (Muchadeyi et al. 2008; Razafindraibe et al. 2008; Mtileni et al. 2011a; Mwacharo et al. 2011) and displayed using 6 pie charts. Assuming that individuals bearing the clade E have originated and dispersed out of the Indian subcontinent following prehistoric seafaring trade, we tested if genetic distance correlates with geographic distance from the population source (India) using Mantel test, as implemented in GenoDive software v2.0b (Meirmans and Tienderen 2004), and the results significance, $z_{xy} = \sum_{i=1}^{n} \sum_{j=1}^{t} x_{ij} y_{ij}$

was tested by permutation (1000 permutations). In this analysis we have gathered all sequences that belong to clade E available at GenBank from India, Kenya, Sudan, and Zimbabwe. Details about the sequence sources, numbers and their accession numbers used for Mantel test are given in Table S4.3. The sequences were grouped into six clusters based on the geographical locations of their corresponding origin as follows: India (the center of origin), Northeast Arabian Peninsula (Oman and Saudi Arabia), Southeast Arabian Peninsula (south Oman, Yemen and Socotra Island), Northeast Africa (Somalia and Sudan), East Africa (Kenya) and Southeast Africa (Zimbabwe). Six coastal points representing prehistoric harbors/cities were selected along the Indian Ocean rim for estimating the physical distances between populations. The geographic distances were obtained based on the great-circle distances method, that is the shortest distance between two points on the surface of a sphere, measured along the surface of the sphere (as opposed to a straight line through the sphere's interior) (http://www.movable-type.co.uk/scripts/latlong.html).

4.3 Results

4.3.1 Within- and between- population diversities

The haplotype diversity ranged from 0.4760 ± 0.1550 (HAF) to 0.8087 ± 0.0360 (CTA) (Table 4.1). On the other hand, the highest nucleotide diversity was observed in HAF (0.0063 ± 0.0040) and the lowest value in NEA (0.0034 ± 0.0023). Across groups, the mean haplotype diversity was 0.7588 ± 0.0300 whereas nucleotide diversity was 0.0065 ± 0.0033.

Table 4.1 Number of haplotypes, haplotype and nucleotide diversity and their standard error for five local chicken groups based on D-Loop mtDNA; Northeast Arabian Peninsula (NEA), Southeast Arabian Peninsula (SEA), Central Arabia (CTA), Socotra Island (SOC) and Horn of Africa (HAF).

Population Name	Sample size	No. of Polymorphic sites	No. of haplotypes	Haplotype diversity *(H)*	Nucleotide diversity (π)
NEA	83	17	12	0.5360 ± 0.0650	0.0034 ± 0.0023
SEA	33	13	7	0.7940 ± 0.0440	0.0058 ± 0.0036
CTA	61	18	13	0.8087 ± 0.0360	0.0047 ± 0.0030
SOC	15	9	4	0.6380 ± 0.0930	0.0037 ± 0.0026
HAF	15	15	5	0.4760 ± 0.1550	0.0063 ± 0.0040
Total/average	207			0.7588 ± 0.0300	0.0065 ± 0.0033

Among groups, the genetic differentiation using population pairwise (Φ_{ST}) (Table 4.2) showed that highest divergence existed between HAF and NEA (0.7848), whereas the lowest was between SOC and SEA (-0.0177). All pairwise comparisons showed significant differences ($P \leq 0.05$), except that between SEA and SOC. The 207 sequences from Arabian chicken (including 175 sequences of our study and 32 from GenBank) generated 27 haplotypes that were defined by 31 variable sites (Table 4.3).

Table 4.2 Population pairwise (Φ_{ST}) between local chicken groups based on D-Loop mtDNA; Northeast Arabian Peninsula (NEA), Southeast Arabian Peninsula (SEA), Central Arabia (CTA), Socotra Island (SOC) and Horn of Africa (HAF).

	NEA	SEA	CTA	SOC	HAF
NEA					
SEA	0.0715^{***}				
CTA	0.0982^{***}	0.0808^{***}			
SOC	0.0563^{*}	-0.0177	0.1255^{**}		
HAF	0.7848^{***}	0.6666^{***}	0.7462^{***}	0.7224^{***}	

Significant differences at *P < 0.05, **P < 0.01 and ***P < 0.001.

Table 4.3 Nucleotide variation found in the 27 haplotypes derived from 207 sequencing data. The number of individuals sharing the same haplotype in each group is presented in the right column. Nucleotide site positioning is relative to the White Leghorn reference sequence (GenBank accession no. X52392; Desjardins and Morais, 1990). Dots (.) denote identity with the reference sequence. N represents total number of individuals in each haplotype.

	130	164	190	196	207	209	214	219	222	230	239	243	246	253	258	278	292	297	303	306+1	307	312	319	327	339	350	352	359	360	414	443	NEA	SEA	CTA	SOC	HAF	N
Ref (X52392)	T	C	C	T	T	G	T	A	T	C	T	C	A	T	C	A	T	C	T	-	C	C	T	C	A	A	T	C	C	C	C						
Clade E																																					
ARE1	.	T	.	.	C	.	C	.	C	.	C	.	.	C	T	.	.	.	.	-	T	.	.	.	.	.	.	.	.	.	T	55	12	22	8		97
ARE2	.	T	.	.	C	.	C	.	C	.	C	.	.	C	T	.	.	.	.	-	T	.	.	.	.	.	.	.	.	T	T	1			1		2
ARE3	.	T	.	.	C	.	C	.	C	.	C	.	.	C	T	.	.	.	.	-	T	.	.	T	.	.	.	.	.	.	T			1		1	2
ARE4	.	T	.	.	C	.	C	G	C	.	C	.	.	C	T	.	.	.	.	-	T	.	.	T	.	.	.	.	.	.	T	2	4	14			20
ARE5	C	T	.	.	C	.	C	G	C	.	C	.	.	C	T	.	.	.	.	-	T	.	.	T	.	.	.	.	.	.	T			2			2
ARE6	.	T	.	.	C	.	C	G	C	.	C	.	.	C	T	.	.	T	.	-	T	.	.	T	.	.	.	.	.	.	T			3			3
ARE7	.	T	.	.	C	.	C	.	C	.	C	.	.	C	T	.	.	.	.	A	T	.	.	.	.	.	.	.	.	.	T	1					1
ARE8	.	T	.	.	C	.	C	.	C	.	C	.	.	C	T	.	.	.	.	-	T	.	.	.	.	G	.	.	.	.	T			4			4
ARE9	.	T	.	.	C	.	C	.	C	.	C	.	.	C	T	.	C	.	.	-	T	.	.	.	.	.	.	.	.	.	T	1					1
ARE10	.	T	.	.	C	.	C	.	C	.	C	.	.	C	T	.	.	.	.	-	T	.	.	.	G	.	.	.	.	.	T	3	4	2			9
ARE11	.	T	A	.	C	.	C	.	C	.	C	.	.	C	T	.	.	.	.	-	T	.	.	.	.	.	.	.	.	.	T	4		1			5
ARE12	.	T	.	.	C	.	C	.	C	.	C	.	.	C	T	.	.	.	.	-	T	.	A	.	.	.	.	.	.	.	T	1					1
ARE13	.	T	.	.	C	.	C	.	C	.	C	T	.	C	T	.	.	.	.	-	T	.	.	.	.	.	.	.	.	.	T	3	8		5		16
ARE14	.	T	.	.	C	.	C	.	C	T	C	T	.	C	T	.	.	.	.	-	T	.	.	.	.	.	.	.	.	.	T			1			1
ARE15	.	T	.	.	C	.	C	.	C	.	C	T	.	.	T	.	.	.	.	-	T	.	.	.	.	.	.	.	.	.	T		1				1
ARE16	.	T	.	C	C	.	C	.	C	.	C	.	.	.	T	.	.	.	.	-	T	.	.	.	.	.	.	.	.	.	T			2			2
ARE17	.	T	.	.	C	.	C	.	C	.	C	.	G	.	T	.	.	.	.	-	T	.	.	.	.	.	C	.	.	.	T		1				1
ARE18	.	T	.	.	C	.	C	.	C	.	C	.	.	A	T	.	.	.	.	-	T	.	.	.	.	.	.	.	.	.	T	7		2			9
ARE19	.	T	.	.	C	.	C	.	C	.	.	.	.	C	T	.	.	.	.	-	T	.	.	.	.	.	.	.	.	.	T			6			6
ARE20	.	T	.	.	C	.	C	.	C	.	C	.	.	C	.	.	.	.	.	-	T	.	.	.	.	.	.	.	.	.	T	1					1
Clade C																																					
ARC1	.	T	.	.	C	.	.	.	C	.	C	.	.	C	.	.	G	.	C	-	T	.	.	.	G	.	.	T	T	.	.					11	11
ARC2	.	T	.	.	C	.	.	.	C	.	C	.	.	C	T	.	G	.	C	-	T	.	.	.	G	.	.	T	T	.	.		3		1		4
ARC3	.	T	.	.	C	.	.	.	C	.	C	.	.	C	.	.	G	.	C	-	T	.	.	.	.	.	.	T	T	.	.					1	1
ARC4	.	T	.	.	C	.	.	.	C	T	C	.	.	C	.	.	G	.	C	-	T	.	.	.	G	.	.	T	T	.	.					1	1
Clade A																																					
ARA1	.	.	.	.	C	.	.	.	.	.	.	.	.	.	.	.	.	.	.	-	.	.	.	.	.	.	.	.	.	.	.					1	1
ARA2	.	.	.	.	.	.	.	.	.	.	.	.	.	.	.	.	.	.	.	-	.	.	.	.	.	.	.	.	.	.	.	4					4
ARB1	.	T	.	.	C	A	.	.	C	.	.	T	.	.	.	.	.	.	.	-	.	T	.	.	.	.	.	.	.	.	.			1			1

4.3.2 Phylogenetic analyses and haplotypes distribution

The haplotypes grouping using median-joining tree method, revealed three distinct clades which we identified as E, A and C (Figure 4.2 and Table S4.2) following Liu's nomenclature (Liu et al. 2006). The individual codes were abbreviated as AR (Arabia) followed by the clade where it belongs (E/A/C) and a number. Specifically, ARE code was given to all haplotypes clustered with or centered on E1 whereas code ARA was given to haplotypes clustered with or centered on A1 and clade B1. The code ARC was given to all haplotypes clustered adjacent to C1 and D1.

Clade E was the most frequent clade with a total of 20 haplotypes (ARE1-ARE20) represented by 183 individuals scattered across the Arabian Peninsula, Socotra Island and in one individual in Somalia (HAF). ARE1, which centered on haplotype E1, was found in 66.3%, 19.7%, 66.7% and 53.3% of NEA, SEA, CTA and SOC, respectively. Other frequent haplotypes in clade E were ARE4, which was found in 20 individuals (CTA, 14; SEA, 4 and NEA, 2) and ARE13, which was observed in 16 individuals (NEA, 13; SEA, 8 and SOC, 5).

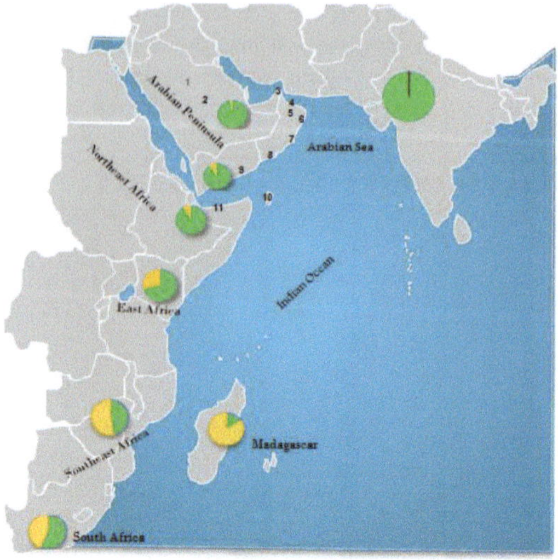

Figure 4.1 Pie charts with the proportion of the Indian (*clade E,* green color) origin and Asian (*clades A, B, C, D;* yellow color) origin in the total of all mitochondrial clades observed across the Indian Ocean rim; The black numbers locate the centroid of sampling per each region: Saudi Arabia [Qassim (1), Riyadh (2)]; Sultanate of Oman [Musandam (3), Batinah (4), North Hajar (5), East Hajar (6), East Coast (7), Dhofar (8)]; Yemen [Mukalla (9), Socotra Island (10)]; Somalia [Hargeysa (11)].

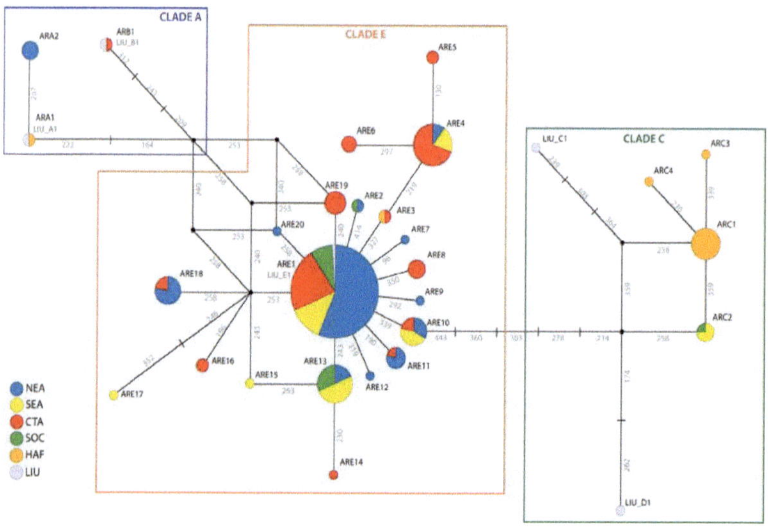

Figure 4.2 Median-joining network of mtDNA D-loop haplotypes observed in Arabian Peninsula, Socotra Island and Somalia chickens as well as most frequent haplotypes reported by Liu et al. (2006). The circle sizes are proportional of the haplotype frequencies. Black circles represent median vectors those connecting indirectly related haplotypes. The numbers on the line correspond to mutational positions connecting haplotypes.

The second most frequent clade was clade C, with a total of 17 chickens representing 4 haplotypes (Table 4.3 and Figure 4.2). Haplotype ARC1 was the major haplotype with a total of 11 individuals from HAF and separated from C1 by 5 mutations. On the other hand, ARC2 was separated from D1 by four mutations and was found in three individuals from SEA and one individual from SOC. Clade A was the least frequent clade in our study and was composed of three haplotypes found in six individuals. Within this clade, haplotype ARA2 had highest frequency and was composed of four individuals from NEA whereas haplotypes ARA1 and ARB1 centered on A1 and B1 and consisted of a single chicken from HAF and CTA, respectively.

4.3.3 Out-of-India

The frequency estimation in terms of clades showed that clade E was the most dominant clade on the Arabian Peninsula and all around East and South Africa (Muchadeyi et al. 2008; Razafindraibe et al. 2008; Mtileni et al. 2011a; Mwacharo et al. 2011) (Figure 4.1). However, the frequency of this clade decreased southward from Arabia to South Africa. For example, clade E was observed in 97.4% of samples from Northeast Arabian Peninsula and slightly decreased to 90.0% and 91.0% in Southeast Arabian Peninsula and Northeast Africa, respectively. The figure dropped to 71% in East Africa, 38.1% in Southeast Africa, and increased again in South Africa (56.8%). Finally, in Madagascar, the frequency of clade E dropped sharply to 16.0%. The Mantel test showed significant (*P=0.02*) and positive correlation (*Mantel r* =0.603) between geographic (km) and genetic distances ($\Phi_{ST}/(1-\Phi_{ST})$) matrices as the genetic distance increased with the geographical distance (Figure 4.3).

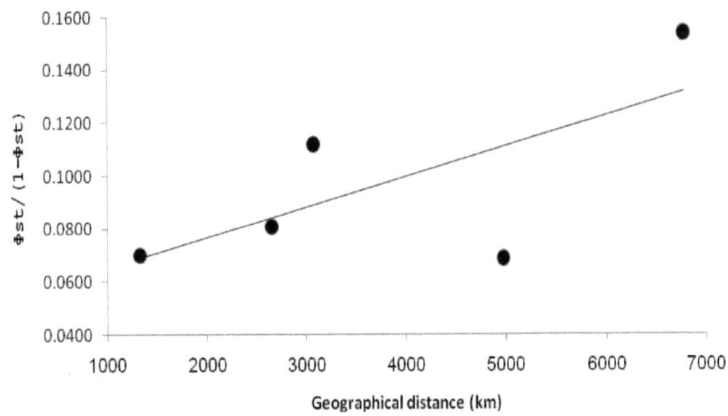

Figure 4.3 Graphic plot of the regression analysis between geographic distance (km) and genetic distance $[\Phi_{ST}/(1-\Phi_{ST})]$ between six regions located around Indian Ocean rim.

4.4 Discussion

Local Arabian chicken displayed relatively higher haplotype diversity similar with those reported for African populations (Muchadeyi et al. 2008; Mtileni et al. 2011a; Mwacharo et al. 2011).

However, the high haplotype diversity observed in Arabian Peninsula chicken mostly falls inside one single clade (E). The high frequency of this clade on the Arabian Peninsula is not surprising, as it originated at the domestication center located on the Indian subcontinent. Also, the higher clade E frequency among other clades in Arabian Peninsula chicken, when compared to that observed in across East Africa countries (Figure 4.1), supports the influence of the Indian subcontinent agricultural center on this area.

The close proximity of the Arabian Peninsula to the Indian subcontinent, particularly to the Indus valley, is a strong argument that found support from many scholars (Edens 1992; Boivin and Fuller 2009; Fuller et al. 2011). It is well documented that the Harappan civilization in the Indus Valley was involved in maritime trade with the Arabian Peninsula, especially Oman and Bahrain, via the Arabian Sea and the Indian Ocean in the second half of the 3rd millennium (Ray 2003; Boivin et al. 2010). Other scholars, however, defend that the much earlier maritime oriented fishing cultures that appeared along the coast of Oman as early as the 7th millennium BP (Biagi 2006), were active players in the trade and dispersal of crops through the coastal areas (Haaland 2011). Very interestingly, recent discoveries of ancient chicken bones in Lothal, on the west coast of India, raised the possibility that the birds could have been carried to the Arabian Peninsula and Mesopotamia as cargo or provision on ships in the third millennium BC (Adler and Lawler 2012). Yet, another genetic study on local Omani cattle (Mahgoub et al. 2013), found a significant contribution of the Indian cattle (Zebu) to the genetic makeup of local cattle populations in Oman. This adds more evidences to the idea that the Arabian Peninsula served as an advanced outpost in the spread of Indus valley domesticated species across the Indian Ocean. Similarly, Fuller and Boivin (2009) pointed to the existence of identical domesticated species in African and Indian savannahs several centuries prior to their binary maritime contact.

Concerning the history of chicken, it has been for long defended that coastal maritime trading networks around the Indian Ocean were the main responsible for the introduction of chicken into Eastern Africa (Williamson 2000; Blench 2003; Muchadeyi et al. 2008; Mtileni et al. 2011b; Mwacharo et al. 2011). Our results from isolation-by-distance hypothesis test demonstrated that chicken populations gained more genetic distinctiveness as the geographical distance from their center of domestication increased (Figure 4.3). The correlation between genetic and geographic distance indicated that nearly 36.4% (r^2=0.364; $P<0.05$) of the variation in genetic distances could be attributed to geographic distances from India following a maritime coastal dispersal route from India to Africa trough the Arabian Peninsula. According to our results, chickens of clade E on the Arabian Peninsula were genetically closer related to their counterparts on the Indian subcontinent than those in East and Southeast Africa. This result indicates that the Arabian Peninsula might have been involved in the early chicken dispersal and represents a historic stopover station in their long distribution route from the center of origin to Africa.

The highest estimated diversity indices suggest that Horn of Africa chicken (HAF) are associated to the high frequency prevalence of clade C, while this clade was absent or observed at low frequencies in the other populations around the upper region of the Indian Ocean rim. The presence of clades C and D in chicken from the Horn of Africa and their absence in commercial and European local chickens (Muchadeyi et al. 2008), may indicate an older introduction of these lineages to this region. Liu et al. (2006) have suggested that clades C and D originated from within a small geographical range in Southwest China and/or surrounding regions such as Vietnam, Burma, Thailand and India. The noticeable presence of clade C and D in the Horn of Africa is an exception from the widespread distribution of clade E in the region. Currently, there are two possible explanations for the presence of these Southeast Asia clades in Africa. One, less plausible however, is related to the Chinese maritime expeditions to the Horn of Africa during the fifteenth century AD (Beaujard 2005). There are historical records indicating that the Chinese emissaries have exchanged gifts with the Somalia rulers, and some of those gifts where animals (Duyvendak 1939). The other explanation might be a secondary expansion of Austronesian chickens throughout Madagascar (Figure 4.1). There is a large body of evidence that Madagascar was colonized by long distance migrations of people from Indonesian Islands (Matthew et al. 2005). Recently, a study on genetics of Madagascar chicken showed that the majority of their chicken carries the Southeast Asia clades and the presence of such clade was associated to the legacy of that long distance maritime migration into Madagascar (Razafindraibe et al. 2008; Mwacharo et al. 2013). Thus, it is plausible that Madagascar has played the same role as the Arabian Peninsula as an intermediary in the introduction of chicken into continental Africa. Interestingly, the frequency and diversity of the clade that originated in Indian decreases from the Arabian Peninsula southwards to Southeast Africa, while the frequency of the Southeast Asiatic clades decreased from Madagascar Northward to East Africa (Figure 4.1).

Finally, concerning the presence of clade A, which was assigned to Yunnan province in China and surrounding areas (Liu et al. 2006), and which is frequent in European chicken (Muchadeyi et al. 2008), probably is due to recent introduction of commercial broilers, purebred brown and white egg-layers. The presence of these two clades in few numbers (6 individuals) could therefore represent signatures of recent introgression of commercial chicken mtDNA haplotypes into village chickens.

4.5 Conclusions

Our study provides additional support to the role of the Indian Ocean in the prehistoric contact between India, the Arabian Peninsula, and Africa. This study may provide the basis for future genetic and archaeological investigations concerning the history of domestication and distribution chickens and other livestock species on the Arabian Peninsula.

Acknowledgements

We would like to thank all farmers who participated in our study. Special thanks go to all persons who provided valuable assistance during sampling: Dr. R. Aljumaah and Dr. R. Atiyat (Saudi Arabia); M. Al-Hashmi (Somalia); M. Alqadhi (Yemen); S. Alhuraizi (Socotra Island). We would like to extend our thanks to Mr. H. Al-Shikaili and Mrs. F. Al-Salmi (Oman) for lab assistance. This work was funded through the Project on "Characterization, Evaluation and

Conservation of Indigenous Animal Genetic Resources in the Sultanate of Oman" (HM Fund SR/AGR/ANVS/08/01) and through a Portuguese Science Foundation (FCT) research grant (PTDC/CVT/117851/2010 and FCOMP-01-0124-FEDER-020141). The last author is supported by the program Science 2008 from FCT.

References

AbdulNayeem M. (2000) *The rock art of Arabia: Saudi Arabia, Oman, Qatar, the Emirates and Yemen*. Hyderabad Publishers, Hyderabad, India.

Adler J. and Lawler A. (2012) How the chicken conquered the world. *Smithsonian* **43**, 40-47.

Al-Abri A.R., Al-Rawas O., Al-Yahyaee S., Al-Habori M., Al-Zubairi A.S. and Bayoumi R. (2012) Distribution of the lactase persistence-associated variant alleles -13910*T and -13915*G among the people of Oman and Yemen. *Human Biology* **84**, 271-286.

Al-Qamashoui B., Mahgoub O., Kadim I. and Schlecht E. (2014) Towards conservation of Omani local chicken: phenotypic characteristics, management practices and performance traits. *Asian-Australasian Journal of Animal Sciences* **27** (6), 767-777.

Al-Yousef Y.M. (2007) A survey study on the distribution of Saudi Baladi chickens and their characteristics. *International Journal of Poultry Science* **6**, 289-292.

Badro D.A., Douaihy B., Haber M., Youhanna S.C., Salloum A., Ghassibe-Sabbagh M., Johnsrud B., Khazen G., Matisoo-Smith E., Soria-Hernanz D.F., Wells R.S., Tyler-Smith C., Platt D.E., Zalloua P.A. and The Genographic Consortium (2013) Y-Chromosome and mtDNA genetics reveal significant contrasts in affinities of modern Middle Eastern populations with European and African populations. *PLoS ONE* 8, e54616.

Bandelt H.J., Forster P., Sykes B.C. and Richards M.B. (1995) Mitochondrial portraits of human populations using median networks. *Genetics* **141**, 743-753.

Beaujard P. (2005) The Indian Ocean in Eurasian and African world-systems before the sixteenth century. *Journal of World History* **16**, 411-465.

Biagi P. (2006) The shell-middens of the Arabian Sea and Persian Gulf. maritime connections in the 7th millennium BP. *Adumatu* **14**, 7-16.

Blench R.M. (2003) The movement of cultivated plants between Africa and India in prehistory. In: *Food, fuel and fields: Progress in African archaeobotany*. Neumann K.A. and Kahlheber S. (Eds), pp. 273–292. Heinrich Barth Institut, Köln.

Boivin N., Blench R. and Fuller D.Q. (2010) Archaeological, linguistic and historical sources on ancient seafaring: a multidisciplinary approach to the study of early maritime contact and exchange in the Arabian Peninsula. In: *The evolution of human populations in Arabia: vertebrate paleobiology and paleoanthroplogy* Petraglia M.D. and Rose J.I. (Eds), pp. 251-278. Springer Netherlands, UK.

Boivin N. and Fuller D.Q. (2009) Shell middens, ships and seeds: exploring coastal subsistence, maritime trade and the dispersal of domesticates in and around the ancient Arabian Peninsula. *Journal of World Prehistory* **22**, 113-180.

Dana N., Megens H.J., Crooijmans R.P., Hanotte O., Mwacharo J., Groenen M.A. and Van Arendonk J.A. (2010) East Asian contributions to Dutch traditional and western commercial chickens inferred from mtDNA analysis. *Animal Genetics* **42**, 125-133.

Duyvendak J.J.L. (1939) *The true dates of the Chinese maritime expeditions in the early fifteenth century*. Brill, The Netherlands.

Edens C. (1992) Dynamics of trade in the ancient Mesopotamian "world system". *American Anthropologist* **94**, 118-139.

Excoffier L. and Lischer H.E.L. (2010) Arlequin suite ver 3.5: A new series of programs to perform population genetics analyses under Linux and Windows. *Molecular Ecology Resources* **10**, 564-567.

Fuller D.Q., Boivin D., Hoogervorst T. and Allaby R. (2011) Across the Indian Ocean: the prehistoric movement of plants and animals. *Antiquity* **85**, 544–558.

Fuller D.Q. and Boivin N.L. (2009) Crops, cattle and commensals across the Indian Ocean: current and potential archaeobiological evidence. *Etudes Ocean Indien* **42/43**, 13–46.

Fumihito A., Miyake T., Takada M., Shingu R., Endo T. and Gojobori T. (1996) Monophyletic origin and unique dispersal patterns of domestic fowls. *Proceedings of the National Academy of Sciences of the United States of America* **93**, 6792-6799.

Galtier N., Nabholz B., Glemin S. and Hurst G.D. (2009) Mitochondrial DNA as a marker of molecular diversity: a reappraisal. *Molecular Ecology* **18**, 4541-4550.

Gongora J., Rawlence N.J., Mobegi V.A., Jianlin H., Alcalde J.A., Matus J.T., Hanotte O., Moran C., Austin J.J., Ulm S., Anderson A.J., Larson G. and Cooper A. (2008) Indo-European and Asian origins for Chilean and Pacific chickens revealed by mtDNA. *PNAS* **105**, 10308–10313.

Groucutt H.S. and Petraglia M.D. (2012) The prehistory of the Arabian Peninsula: deserts, dispersals, and demography. *Evolutionary Anthropology* **21** (3), 113-125.

Haaland R. (2011) Crops and Culture: dispersal of African millets to the Indian subcontinent and its cultural consequences. *Dhaulagiri Journal of Sociology and Anthropology* **5**, 1-30.

Kadim I., Al-Marzooqi W., Al-Waheebi S. and Mahgoub O. (2009) Improving egg and meat production of Omani local chicken through genetic and phenotypic selection programmes. Report, Sultan Qaboos University, Muscat, Oman.

Kanginakudru S., Metta M., Jakati R.D. and Nagaraju J. (2008) Genetic evidence from Indian red jungle fowl corroborates multiple domestication of modern day chicken. *BMC Evolution Biology* **8**, 174.

Kryger K.N., Thomsen K.A., Whyte M.A. and Dissing M. (2010) Smallholder poultry production; livelihoods, food security and sociocultural significance. FAO(Ed), Rome, Italy.

Liu Y.P., Wu G.S., Yao Y.G., Miao Y.W., Luikart G., Baig M., Beja-Pereira A., Ding Z.L., Palanichamy M.G. and Zhang Y.P. (2006) Multiple maternal origins of chickens: out of the Asian jungles. *Molecular Phylogenetics and Evolution* **38**, 12-19.

Mahgoub O., Babiker H.A, Kadim I.T., Al-Kindi M., Hassan S., Al-Marzooqi W., Eltahir Y.E., Al-Abri M.A., Al-Khayat A., Al-Sinani K.R., Al-Khanjari H., Costa V., Chen S. and Beja-Pereira A. (2013) Disclosing the origin and diversity of Omani cattle. *Animal Genetics* **44**, 336-339.

Matthew E.H., Bryan C.S., Mark A.J. and Peter F. (2005) The dual origin of the Malagasy in Island Southeast Asia and East Africa: evidence from maternal and paternal lineages. *American Journal of Human Genetics* **76** (5), 894–901.

Meirmans P.G. and Tienderen P.H. (2004) GENOTYPE and GENODIVE: two programs for the analysis of genetic diversity of asexual organisms. *Molecular Ecology Notes* **4**, 792-794.

Miao Y.W., Peng M.S., Wu G.S., Ouyang Y.N., Yang Z.Y., Yu N., Liang J.P., Pianchou G., Beja-Pereira A., Mitra B., Palanichamy M.G., Baig M., Chaudhuri T.K., Shen Y.Y., Kong Q.P., Murphy R.W., Yao Y.G. and Zhang Y.P. (2013) Chicken domestication: an updated perspective based on mitochondrial genomes. *Heredity* **110**, 277–282.

Mtileni B.J., Muchadeyi F.C., Maiwashe A., Chimonyo M., Groeneveld E., Weigend S. and Dzama K. (2011a) Diversity and origin of South African chickens. *Poultry Sciences* **90**, 2189–2194.

Mtileni B.J., Muchadeyi F.C., Maiwashe A., Groeneveld E., Groeneveld L.F., Dzama K. and Weigend S. (2011b) Genetic diversity and conservation of South African indigenous chicken populations. *Journal of Animal Breeding and Genetics* **128**, 209-218.

Muchadeyi F.C., Eding H., Simianer H., Wollny C.B.A., Groeneveld E. and Weigend S. (2008) Mitochondrial DNA D-loop sequences suggest a Southeast Asian and Indian origin of Zimbabwean village chickens. *Animal Genetics* **39**, 615–622.

Mwacharo J.M., Bjørnstad G., Han J.L. and Hanotte O. (2013) The history of African village chickens: an archaeological and molecular perspective. *African Archaeological Review* **30**, 97-114.

Mwacharo J.M., Bjørnstad G., Mobegi V., Nomura K., Hanada H., Amano T., Jianlin H. and Hanotte O. (2011) Mitochondrial DNA reveals multiple introductions of domestic chicken in East Africa. *Molecular Phylogenetics and Evolution* **58**, 374-382.

Nei M. (1978) Estimation of average heterozygosity and genetic distance from a small number of individuals. *Genetics* **89**, 583–590.

Nei M. and Tajima F. (1981) DNA polymorphism detectable by restriction endonucleases. *Genetics* **97**, 145.

Pickering H. (2007) *Frankincense; the perfume of Arabia*. Al Roya Press and Publishing House, Muscat, Sultanate of Oman.

Potts D.T. (2000) *Ancient Magan, the secrets of Tell Abraq*. Trident Press Ltd., UK.

Ray H.P. (2003) *The Archaeology of Seafaring in Ancient South Asia*. Cambridge University Press, UK.

Razafindraibe H., Mobegi V.A., Ommeh S.C., Rakotondravao M.L., Bjørnstad G., Hanotte O. and Jianlin H. (2008) Mitochondrial DNA origin of indigenous Malagasy chicken. *Annals of the New York Academy of Sciences* **1149**, 77-79.

Silva P., Guan X., Ho-Shing O., Jones J., Xu J., Hui D., Notter D. and Smith E. (2009) Mitochondrial DNA-based analysis of genetic variation and relatedness among Sri Lankan indigenous chickens and the Ceylon junglefowl (*Gallus lafayetti*). *Animal Genetics* **40**, 1-9.

Storey A.A., Athens J.S., Bryant D., Carson M. and Emery K. (2012) Investigating the global dispersal of chickens in prehistory using ancient mitochondrial DNA Signatures. *PLoS ONE* **7** (7), e39171.

Tamura K., Peterson D., Peterson N., Stecher G., Nei M. and Kumar S. (2011) MEGA5: Molecular evolutionary genetics analysis using maximum likelihood, evolutionary distance, and maximum parsimony methods. *Molecular Biology and Evolution* **28**, 2731-2739.

Williamson K. (2000) Did chicken go west? In: *The origins and development of African livestock: Archaeology, genetics, linguistics and ethnography.* Blench R.M. and MacDonald K.C. (Eds), pp. 368–448. UCL Press, London, UK.

Wragg D., Mwacharo J.M., Alcalde J.A., Hocking P.M. and Hanotte O. (2012) Analysis of genome-wide structure, diversity and fine mapping of Mendelian traits in traditional and village chickens. *Heredity* **109** (1), 6-18.

Supplementary material

Table S4.1 Distribution of haplotypes in our study groups and corresponding populations. Chicken population abbreviations: MU = Musandam; BT = Batinah; NH = North Hajar; EH = East Hajar; EC = East Coast; DF = Dhofar; MK = Mukalla; RY = Riyadh; QS = Qassim; HD = Hadibo; HR = Hargeysa.

	Northeast Arabia (NEA)					Southeast Arabia (SEA)		Central Arabia (CTA)		Socotra Island (SOC)	Somalia (HAF)
clade E	MU	BT	NH	EH	EC	DF	MK	RY	QS *	HD	HR
ARE1	9	11	11	15	9	5	7	21	1	8	
ARE2	1									1	
ARE3									1		1
ARE4	1	1				4			14		
ARE5									2		
ARE6									3		
ARE7	1										
ARE8								4			
ARE9				1							
ARE10	2	1				4			2		
ARE11	1	1			2			1			
ARE12					1						
ARE13			1	1	1	2	6			5	
ARE14								1			
ARE15							1				
ARE16									2		
ARE17							1				
ARE18		1	3		3			2			
ARE19									6		
ARE20			1								
clade C											
ARC1											11
ARC2						2	1			1	
ARC3											1
ARC4											1
clade A											
ARA1											1
ARA2	3		1								
ARB1									1		
Total	18	15	17	17	16	17	16	29	32	15	15

* Sequences were retrieved from a study of Yacoub and Fathi (2013).

Table S4.2 Nomenclature of chicken clades in our study and correspondence with other studies; NEA (Northeast Arabian Peninsula), CTA (Central Arabia), SOC (Socotra Island), SEA (Southeast Arabian Peninsula) and HAF (Horn of Africa)

This study	Liu et al. (2006)	Mwacharo et al. (2011)	Muchadeyi et al. (2008)	Mtileni et al. (2011a)
Clade E Common in NEA, SEA, CTA, SOC and in one individual in HAF	Clade E Observed in Europe, Middle East and India. Possible postulated center of origin: Indian subcontinent.	Clade D Most widespread in East Africa, Ethiopia, Uganda and Sudan.	Clade C Observed in Zimbabwe and Sudan village chickens.	Clade E Observed in conserved and field chickens of South Africa.
Clade C Common in Somalia, few in SEA and SOC.	Clade C Observed in Guangxi and Guangdong China and Japan. Clade D Observed in Indonesia and India, and in Chinese and Japanese gamecocks. Possible recent domestication in South and Southwest China and/or surrounding areas (i.e. Vietnam, Burma, Thailand, India.	Clades A Observed only in Kenya, mostly around the coastal regions.	Clade A Observed in Zimbabwe, Malawi and Sudan village chickens.	
Clade A Observed mainly in NEA.	Clade A Observed in South China and Japan. Possible postulated center of origin: Yunnan province and/or surrounding areas. Clade B Observed in Yunnan China. Suggested center of origin: Yunnan China and/or surrounding areas.	Clades B and C Observed in a single individual in Kenya, mostly around the coastal regions and in Ethiopia.	Clade B1 Frequent in purebred brown and white egg layers. Clade B2 Common in commercial broilers and chicken from Northwest Europe.	Clade A Found in conserved and field chickens of South Africa.

Table S4.3 Description of 548 mtDNA sequences clustered with clade E used in Mantel test analyses.

Region	No. of sequences	Retrieved haplotypes	Accession numbers	Source study
India	247	45	EU847802, EU847804-EU847816, GU447485, GU447490, GU447492, GU447495, GU447581, GU447585-GU447589, GU448259, GU448269, GU448271, GU448272, GU448275, GU448357-GU448373, GU448375-GU448386, GU448388, GU448389, GU448391-GU448394, GU448396-GU448399, GU448401-GU448404, GU448407, GU448409-GU448418, GU448420, GU448421, GU448423, GU448425, GU448426, GU448428-GU448430, GU448432-GU448453, GU448455-GU448465, GU448468-GU448474, GU448477-GU448479, GU448481, GU448483-GU448486, GU448727-GU448729, GU448731-GU448735, GU448737-GU448744, GU448747-GU448749, GU448755-GU448764, GU448766, GU448770-GU448774, GU448779, GU448917-GU448919,GU902213, GU448948-GU448950, GU448952-GU448956, GU448958-GU448967, GU448969, GU557140, GU557142, GU557144, GU557146, GU902198-GU902202,GU902204-GU902206,GU902208-GU902212, GU902214, GU902219, GU902220, GU902222-GU902225, GU902227-GU902229, GU902233-GU902238, GU902240-GU902242, GU902244, AY644966-AY644969, AY644973, AY704702, AY704703, AY704705-AY704708, AY704712-AY704715	(Liu et al. 2006; Kanginakudru et al. 2008; Arora et al. 2010; Singh and Kumar 2011; Miao et al. 2013)
Northeast Arabian Peninsula (North Oman, Saudi Arabia)	139	18	ARE1-ARE14, ARE16, ARE18-ARE20, KC436009-KC436022, KC436024-KC436040	Yacoub and Fathi (2013) and this study
Southeast Arabian Peninsula (South Oman, Yemen, Socotra Island)	44	7	ARE1-ARE2, ARE4, ARE10, ARE13, ARE15, ARE17	This study
Northeast Africa (Somalia and Sudan)	4	4	ARE3, AM746042, AM746045 and AM746046	Muchadeyi et al. (2008) and this study
East Africa (Kenya)	107	21	EU095035, EU095036, EU095038-EU095043, EU095046, EU095048-EU095050, EU095057-EU095064, EU095066, EU095074, EU095075, EU095081, EU095083, EU095085, EU095089, EU095091, EU095093, EU095095, EU095101, EU095102, EU095105, EU095107, EU095109, EU095112, EU095117, EU095120-EU095155, EU095157-EU095181, EU095183-EU095185, EU095187-EU095192	Mwacharo et al. (2011)
Southeast Africa (Zimbabwe)	7	7	AM746031, AM746032, AM746040-AM746044	Muchadeyi et al. (2008)
Total	548	97		

CHAPTER 5

General discussion

5.1 General discussion

Characterization of farm animal genetic resources is a prerequisite for any conservation and sustainable utilization of these resources. A good description and understanding of the production systems of local livestock breeds is required to implement appropriate strategies for improving their production and ensure the involvement of animal owners in conservation programs (FAO 2008). To date, no studies have been carried out to characterize the production and genetic potential of local chickens in Oman. However, some reports documented an increased interest in their social and economic importance (Saleh 2000; Kadim et al. 2009; MAF 2013). The current study aimed to assess the production system, performance and genetic diversity of local chickens in Oman as a prerequisite towards designing appropriate plans for their conservation and improvement.

The main objectives of Chapter 2 were to characterize the production system and to assess the production traits and phenotypic features of local chicken in six agro-ecological zones of Oman. The study revealed a bigger role of women in comparison to men in local chicken husbandry in Oman. In all study areas, women took care of the major daily tasks of the birds such as feeding, cleaning and collecting eggs (Table 2.2) which keep the flock under daily observation. This may explain why flock size was bigger when owned by women as shown by the multiple regression model (Table 2.8). This active participation of women in chicken husbandry indicates that they should be strongly considered in chicken conservation and improvement programs in the future. Indeed, the fact that 65.6% of women in Oman are involved in agricultural activities (MAF 2013) has been positively exploited by decision makers to accomplish several animal husbandry programs targeting rural families in the Sultanate - among these, the extension program for Small-scale Local Chicken Units (SLCU) that has been recently introduced by the Directorate of Rural Women Development of the Ministry of Agriculture and Fisheries(MAF 2013) (see General Introduction).

The production in all AEZ was dominated by free-range scavenging system (Table 2.4). As in many tropical and subtropical countries (IAEA 2004; ACIAR 2005; Pica-Ciamarra and Dhawan 2010), birds under this system were released during daytime to scavenge agricultural by-products and household leftovers. In addition, insufficient housing conditions and management assets were provided (Table 2.4).

Despite the low input and modest requirements, local chicken farming contributed to the income of almost one third of the respondents (Table 2.2). The role of indigenous chicken breeds in securing food and income to rural families has been extensively documented (Kryger et al. 2010). However, agriculture modernization processes in many developing countries forced farmers to accept newly introduced breeds in an effort to improve their productivity and profits (FAO 2004). Consequently, encouraging small-scale farmers to keep and give consideration to the low-productive native breeds may become gradually difficult (Altieri 1999). Therefore, a loss of traditional livestock breeds, either by crossbreeding or by a complete replacement with higher yielding commercial alternatives is expected. Accordingly, a range of essential genetic traits, especially those encoding the adaptation to local conditions slowly, become less frequent in the chicken population (Besbes et al. 2011). For local chickens in Oman, this is particularly important because free scavenging makes chickens susceptible to undesired gene exchange via introgression with commercial birds from neighboring flocks.

Our study revealed several socioeconomic features that efficiently contributed to a better chicken production. Gender, age of household head, knowledge and skills, family size, cropland size, overall livestock endowment, total income, and existence of hired labor (Table 2.5 and Table 2.8) seem to have significant effects on adopting better housing and feeding conditions or to own a larger flock, obtain higher egg production and lower bird mortality. These insights might guide decision makers in the selection process for new candidate farmers to carry out chicken production projects in the future.

Training had a positive effect towards adopting better feeding and housing for local chicken flocks (Table 2.5) emphasizing the important role of governmental extension programs in improving management skills of chicken owners. This possibly has been achieved by improving the basic knowledge and skills of farmers, or through their direct contacts with veterinary services and extension agents. Indeed, training-based programs have been successfully conducted by governments and associations to launch commercially-oriented small-scale local chicken projects (IAEA 2004; Pica-Ciamarra and Dhawan 2010). The Ministry of Agriculture and Fisheries can have a major role in achieving such promising developments by means of extension programs. Women should receive technical training (housing, feeding, health care, and general management of the poultry) in order to be able to generate some income from profitable semi-scavenging poultry.

Omani local chickens possess a wide range of morphological and phenotypic traits that vary within and among the populations in all study areas (Table 2.6 and Table S2.1). The large variation of phenotypic traits is considered as one of the principal features that characterize local chicken production system in tropics and subtropics (Jens et al. 2004). The low priority of color traits in selection strategies of farmers (Table 2.3) and absence of any controlled mating in the scavenging systems might explain the wide range of plumage color revealed in the Omani local chicken. The absence of effects of agro-ecological zone on chicken performance and phenotypic traits (Table 2.6, 2.7 and Table S2.1) indicates that smallholder poultry selection and production strategies were similar across zones.

The main objective of Chapter 3 was to assess the genetic diversity and evaluate the contribution of local populations to the total genetic diversity of Omani chickens for future conservation programs using microsatellite molecular markers. Information from microsatellites can provide reliable estimates of genetic diversity within and between populations and assess conservation priorities among livestock groups. Twenty-nine loci in 158 individuals representing six AEZ chicken populations were genotyped.

A total of 217 alleles were found across 29 microsatellite loci with an average expected heterozygosity of 0.62 across populations (Table 3.1). The figures were comparable with those reported for African village chickens (Muchadeyi 2007; Mtileni et al. 2011_b) and some Asian breeds (e.g. Ha Giang breed in Vietnam; Berthouly et al. (2009). The low F_{ST} (0.034) and positive F_{IS} (0.134) values observed in our study (Table 3.1, 3.2) could be attributed to the counterweight effect of inbreeding and genetic intermixing between chickens due to extensive exchange and movement of genetic stocks among local farmers (Mwacharo et al. 2013).

The neighbor-joining (NJ) tree analysis based on Nei's standard genetic distance showed that all six local populations are clustered independently from commercial chicken clusters, that is Brown egg layers (BL_A, BL_C and BL_D), Broilers (BRD_A, BRD_D, BRS_A and BRS_B) and White egg layers (WL_A and WL_C). This genetic distinction indicates a long time genetic isolation and suggests that local farmers in these AEZ did not introduce any commercial line or breed into their chicken flocks. The NJ tree showed that both wild breeds (RJFG and RJFSC) are quite distant from Omani chickens, therefore recent contribution of *Gallus gallus gallus* from Thailand and *Gallus gallus spadiceus* from China to the Omani chickens is rather unlikely.

Absence of genetic substructuring could be a result of large effective population size (Muchadeyi et al. 2007), high gene flow or a combination of both (Besbes et al. 2011). Ecological factors and production system can lead to large genetic fixation of breeds resulting in genetic substructuring (Leroy et al. 2012). Cuc (2010) observed a clear genetic differentiation among Vietnamese chicken breeds and attributed this to the consequence of isolation and environmental fluctuations between different agro-ecological zones. For the five populations of northern Oman, it is clear that short geographical distances, enhanced by similar production elements, showed no significant effect on their genetic structuring. For DF (in southern Oman), a moderate differentiation from the northern populations was revealed by F_{ST} values and NJ tree analyses (Table 3.2, Figure 3.1). Taking into account that geographical isolation of populations and environmental fluctuations may lead to genetic differentiation, the genetic substructure indicated that local chickens in the northern regions (MU, BT, NH, EH and EC) and in the southern region (DF) may constitute distinct locally-adapted populations. As it is assumed that microsatellites evolve under a nearly neutral model (Oliveira et al. 2006), it is not plausible to conclude that microsatellites are involved in adaptive genetic diversity (Muchadeyi 2007). The limited sample size of the DF population limits full conclusion on genetic structuring patterns.

Genetic and/or phenotypic differentiation into northern and southern breeds has been documented in goat and cattle species in Oman. Al-Araimi (2011) revealed a clear genetic differentiation between northern breeds and the southern (DF) breed of goat based on microsatellites and phenotypic traits. He attributed this to geographical isolation and different origins. Dhofari cattle, similarly, has been found to show different phenotypic and production characteristics than the northern breed (DGALR 2011). However, a genetic diversity assessment is necessary to confirm the latter findings.

In the conservation analyses, population DF showed the highest ranking in the priority for conservation, supported by its higher impact on between-populations genetic diversity (Table 3.3). DF showed also a high contribution to the within-population diversity. Population MU came second in the conservation priority, supported by its high within-population diversity. The importance of DF and MU as valuable elements both of uniqueness and genetic diversity was attributed to their geographic location among the Omani populations. MU is the most northern and DF the most southern population studied (Figure 2.1), therefore, their genetic distinctiveness from other Omani chicken may be attributed to their splitting from the more homogenous gene pool in the central coastal and mountain regions of Oman.

The findings reported in Chapters 2 and 3 present an example of the impact of production and socioeconomic features on the genetic characteristics of local chickens. Free-range scavenging as a major production system in the study are as allowed genetic exchange between chicken flocks and reduced differentiation levels among chicken populations. Absence of selection/breeding schemes has, on the other hand, given their gene pool its wide diversity. In many cases, however, extremely extensive scavenging systems could be unsafe in terms of genetic erosion in the long run. Although the respondents confirmed that their new replacement flocks were of "pure" local breeds and came from well-known sources, an accidental introgression and migration of commercial breeds from neighboring sources might have occurred.

From conservation point of view, maintaining local breeds under their original agro-ecological conditions (i.e. *in vivo /in situ*), where farmers manage their animals in a traditional way, can protect farm animal genetic resources against loss (Gibson et al. 2005). Köhler-Rollefson (1997) stated that for any animal genetic resources, the best way to conserve breeds is by maintaining them as part of functional production systems and in the social and ecological backgrounds in which they were developed. Indeed, a major adverse consequence of local genetic resources loss in the long term could be the undesired transformation of the cheap traditional, agro-ecologically adapted production systems (Altieri 1999). The scavenging system requires low external inputs and capital, provides local chickens with high genetic variation and contributes to their immunity capabilities and adaptation to harsh environmental conditions (Besbes et al. 2011). It is assumed that when raising chickens under such harsh conditions, diverse alleles and allele combinations are produced through natural selection that give these breeds adaptation and a reasonable ability to produce (Muchadeyi 2007). Therefore, any opportunity that can help to conserve these local breeds by combining slightly improved but still cheap management as well as a continuous dynamic adaptation to the environment will be preferable. The SLCU program can serve as a preliminary step towards such combination approach. By minor improvements in management practices, extensive scavenging could be upgraded to a semi-scavenging system. This may efficiently increase the production potential of local chickens and help to preserve them from extinction (Sarkar and Golam 2009).

Very limited information is available about the history of local chickens and their introduction to Oman and the Arabian Peninsula. The mtDNA marker is useful for studying the evolutionary relationships and phylogeny of organisms (Liu et al. 2006). The main aims of Chapter 4 were to assess the population structure and unveil the maternal origins of local chickens on the Arabian Peninsula (Oman, Saudi Arabia, Yemen and Socotra Island) and the Horn of Africa (Somalia) based on mtDNA.

The study revealed 27 haplotypes divided into three distinct clades -clade E, clade C and clade A- in chicken populations of the Arabian Peninsula. Clade E (ARE) was the most common with a total of 20 haplotypes scattered through all study regions and centering on haplotype E1 of Liu (Liu et al. 2006).

With regards to Oman, the 100 mtDNA sequencing data from local chicken populations in six AEZ (locations: 3-8 in Figure 4.1) generated 13 haplotypes (Table 4.S1). The average of haplotype and nucleotide diversities were 0.626 and 0.004, respectively (data are not shown). The majority of chickens in Oman were positioned in clade E (94.9%), which might be the reason for the unstructured populations among AEZ (Figure 4.2). With the exception of two individuals clustered in clade C, population DF shares all its haplotypes in clade E with the northern Omani populations. The fact that the majority of DF chickens share similar maternal lineages with other Omani populations is in contrast to the phylogenetic analyses of microsatellites that showed a clear genetic isolation of DF (Figure 3.1). Clades A and B have appeared in few individuals in MU and NH. Due to sample size variation in the microsatellites and mtDNA studies, it is difficult to interpret the dissimilarity in structuring patterns. In general, it was very clear that the haplotypes-sharing pattern of the six populations was corresponding to the phylogenetic analyses unveiled in Chapter 3. Nevertheless, the current findings confirmed that local field populations possess high genetic diversity and a rich gene pool. This supports the need of conservation for these populations.

According to Liu et al. (2006), clade E supposedly has originated on the Indian subcontinent. The wide presence of clade E in local chickens of the Arabian Peninsula could be a result of a coastal arrival via the Indian Ocean and Arabian Sea, which reflects the special geographical location of Oman and the Arabian Peninsula as a historical melting point and a cross road between major old continents. Given the fact that the Indian subcontinent is the origin of many animal and crop species that have been dispersed to Africa via the Arabian Peninsula for millennia (Boivin and Fuller 2009), it is highly plausible that chicken are a part of these interactions. The dominancy of clade E was reported for many African chicken populations (e.g., Ethiopia, Sudan, Kenya, Uganda, Zimbabwe, Malawi, South Africa; Muchadeyi et al. 2008; Mtileni et al. 2011a; Mwacharo et al. 2011) as illustrated in Figure 4.1.

The findings of this study (Figure 4.3) demonstrated that genetic distances between chicken populations along the Indian Ocean rim could be attributed to geographic distances following a maritime coastal dispersal route from India to Africa trough the Arabian Peninsula. Accordingly, this indicates that the Arabian Peninsula might have been involved in chicken dispersal, representing a historic channel in the domestication and distribution route from their center of origin into Africa. The Arabian Peninsula is discussed as possible introduction point of livestock to Africa along their dispersal route throughout the world. For instance, Rege (1999) and Hanotte et al. (2002) have suggested that Southern Arabia was an introduction point of Indian and Arabian zebu (*Bos indicus*) and the Near Eastern *Bos taurus* to Northeast Africa. Such a dispersal route might have been catalyzed by the trade and cultural connections between settlements in Arabia and the African Horn in the pre-second millennium BC (Potts 2000) and through the Islamic settlements established by Arabs from about the end of the 7th century AD (Epstein 1971; Hourani 1991).

Clade C that has been widely found in Somalian chickens (HAF) originated in Southwest China and/or surrounding regions such as Vietnam, Burma and Thailand (Liu et al.2006). The presence of clades C and D of Liu (Liu et al. 2006) in HAF chickens and their absence in commercial and European local chickens (Muchadeyi et al. 2008) can explain the absence of crossbreeding with commercial chicken lines. Chickens of clades A and B have been domesticated in Yunnan province in China (Liu et al. 2006). Muchadeyi et al. (2008) have observed these two clades in chicken from Northwest Europe, commercial broilers, and purebred brown and white egg layers. Therefore, their presence in local chickens in Oman might represent signatures of recent introgression of commercial chicken mtDNA haplotypes. It is possible that these maternal lineages have been introduced to the country recently by commercial flocks through large-scale poultry producing companies.

5.2 Conclusions

This study is the first to assess the production traits and phenotypic features of Omani local chicken and to investigate their genetic diversity by using molecular techniques. From this study, the following conclusions can be drawn:

1- Despite the variation in farming activities across the major six AEZ of Oman, free-range scavenging is the major production system of local chickens in Oman.
2- The production of local chickens in Oman is under the responsibility of women, and chicken owners in all AEZ display similar production trait priorities for replacement and breeding stock.
3- Taking into account the high genetic and phenotypic diversity besides their capability to convert quantitatively and qualitatively inferior inputs into economical and nutritional benefits for their owners, Omani local chickens can be considered as a rich legacy to the country's genetic resources and as an asset for the future.
4- Socioeconomic conditions and training have been shown to affect some production and performance traits of local chicken in Oman. Thus, considering them in the process of designing chicken improvement programs will be essential.
5- Except for the southern population DF that possesses a unique gene pool, local chicken populations from northern Oman show an absence of substructuring at the autosomal level, indicating gene flow and absence of isolation between populations.

6- Local chicken in Oman and on the Arabian Peninsula seem to derive from three maternal lineages. The dominance of clade E (of Indian maternal lineage) in chickens of Oman and across the rest of the Arabian Peninsula support a coastal introduction scenario. This provides once more evidence for the role of sea trade across the Indian Ocean and Arabian Sea for the prehistoric contact between India, the Arabian Peninsula, and Africa.

5.3 Implications and recommendations

The findings of this study demonstrate the important role of local chicken as a low-input source of food and a source of additional income for smallholder farmers of Oman. This reflects in the governmental decision to initiate programs aiming at improvement and sustainable utilization of these chickens.

Since women are the backbone of local chicken production activities in Oman, giving more attention to them, such as by provision of extensional assets and training in modern chicken husbandry skills, seems of high importance. The productivity of local chickens at farm level could be enhanced by simple changes in management practices such as feeding, housing and health care. Therefore, women should be trained for proper housing, diseases and predator control, in view of reducing chicken mortality. Field workshops could be used to demonstrate utilization of locally available feed ingredients and formulate supplementary rations for chickens. Yet, a study to determine the nutrient composition of the locally available scavenging feeds tuffs should initially be carried out.

Any improvement program should take into consideration the small farmer capacities and capabilities in terms of resources. The already ongoing Small-scale Local Chicken Unit program (SLCU) may present an upgraded form of the low-input scavenging system and provide an opportunity for farmers' involvement in local chicken conservation. A long-term follow-up process for this project is essential to determine its appropriateness to smallholders and success in achieving its goals.

The study also showed that chicken owners do not face difficulties in marketing their chicken products. This suggests that there are preferences of consumers for local chickens. However, from the present data it is not possible to derive demand figures for local chicken products as compared to products from commercial chicken. Therefore, an in-depth social survey is needed to generate more information on this aspect. Moreover, evaluating eggs or meat characteristics such as carcass yield, colour, tenderness, and fatty acid composition of Omani local chickens will be very useful in identifying reasons for such preferences.

The results concerning the genetic makeup of the local chicken populations across Oman point to the existence of two fairly distinct population structures; DF in south and the all other five in the north. All populations possesses high heterozygosity levels and allele numbers, therefore, any conservation strategy must maintain their genetic structure. The populations with higher contribution to both within- and between- genetic diversities (DF and MU) should have higher priority for conservation. The current study may serve as a basis for policy makers and scientists

/ ministry staff to initiate the creation of a nuclear flock for a long-term conservation program in Oman.

The high-density SNPs (single nucleotide polymorphism) array genotyping data available for chicken (e.g. 60K SNP chip) covers the entire genome. Therefore, genome-wide association studies can and should be conducted to identify association between phenotypes and genomic variation. This, associated with other Next Generation sequencing approaches, will allow to identify polymorphisms in regions (e.g. genes) underlying many important phenotypic traits such as production (e.g., meat production and quality, egg production), reproduction (e.g., clutch size and numbers) and adaptive traits (e.g., heat and disease resistance) in Omani local chickens. Landscape genomics approaches would be useful to connect genetic differentiation with environmental conditions.

References

ACIAR (2005) Village chickens, poverty alleviation and the sustainable control of Newcastle disease. In: *Proceedings of an international conference of agricultural research.* Alders R.G., Spradbrow P.B. and Young M.P. (Eds), 5–7 October, Dar es Salaam, Tanzania.

Al-Araimi N. (2011) A study on the genetic diversity of Omani goats using microsatellite DNA markers. *Master Thesis, Department of Animal and Veterinary Sciences, Sultan Qaboos University, Muscat, Oman.*

Altieri A.M. (1999) The ecological role of biodiversity in agroecosystems. *Agriculture, Ecosystems and Environment* **74**, 19–31.

Berthouly C., Leroy G., Nhu Van T., Hoang Thanh H., Bed'Hom B., Trong N.B., Vu Chi C., Monicat F., Tixier-Boichard M., Verrier E., Maillard J.C. and Rognon X. (2009) Genetic analysis of local Vietnamese chickens provides evidence of gene flow from wild to domestic populations. *BMC Genetics* **10** (1).

Besbes B., Tixier-Boichard M., Hoffmann I. and Jain G.L. (2011) Future trends for poultry genetic resources. FAO (Ed), Rome, Italy.

Boivin N. and Fuller D.Q. (2009) Shell middens, ships and seeds: exploring coastal subsistence, maritime trade and the dispersal of domesticates in and around the ancient Arabian Peninsula. *Journal of World Prehistory* **22**, 113-180.

Cuc N.T.K. (2010) Vietnamese local chicken breeds: genetic diversity and prioritising breeds for conservation. *PhD Thesis, Department of Animal Sciences, Georg-August-Universität Göttingen, Germany.*

DGALR (Direcorate General of Agriculture and Livestock Research) (2011) Agriculture and livestock, five-year research strategy (2011-2015). MAF (Ed), Muscat, Oman.

Epstein H. (1971) *The origins of the domestic animals of Africa.* Africana Publishing Corporation, London, UK.

FAO (2004) The state of the world's animal genetic resources for food and agriculture. Rischkowsky B. and Pilling D. (Eds), Rome, Italy.

FAO (2008) Pictorial guidance for phenotypic characterization of chickens and ducks. In: *Manuel Luque Cuesta,* Rome, Italy.

Gibson J., Gamage S., Hanotte O., Iñiguez L., Maillard J.C., Rischkowsky B., Semambo D. and Toll J. (2005) Options and strategies for the conservation of farm animal genetic resources. In: *Report of an International Workshop.* FAO (Ed), Montpellier, France.

Hanotte O., Bradley D.G., Ochieng J.W., Verjee Y., Hill E.W. and Rege J.E.O. (2002) African pastoralism: genetic imprints of origins and migrations. *Science* **296**, 336–339.

Hourani A. (1991) Tribes and states in Islamic history. In: *Tribes and state formation in the Middle East.* Khoury P. and Kostiner K. (Eds), pp. 303–312. IB Tauris, London, UK.

IAEA (2004) Improving farmyard poultry production in Africa: interventions and their economic assessment. In: *Proceedings of a final research coordination meeting,* 24-28/May, Vienna, Austria.

Jens R., Anders P., Charlotte V., Ainsh M. and Lone F. (2004) *Keeping village poultry; A technical manual for small-scale poultry production.* 89 p. Network for Smallholder Poultry Development, Copenhagen, Denmark.

Kadim I., Al-Marzooqi W., Al-Waheebi S. and Mahgoub O. (2009) Improving egg and meat production of Omani local chicken through genetic and phenotypic selection programmes. Sultan Qaboos University, Muscat, Oman.

Köhler-Rollefson I. (1997) Indigenous practices of animal genetic resources management and their relevance for the conservation of domestic animals in developing countries. *Journal of Animal Breeding and Genetics* **114**, 231–238.

Kryger K.N., Thomsen K.A., Whyte M.A. and Dissing M. (2010) Smallholder poultry production – livelihoods, food security and sociocultural significance. FAO (Ed), Rome, Italy.

Leroy G., Kayang B.B., Youssao I.A., Yapi-Gnaore C.V., Osei-Amponsah R., Loukou N.E., Fotsa J.C., Benabdeljelil K., Bed'hom B., Tixier-Boichard M. and Rognon X. (2012) Gene diversity, agroecological structure and introgression patterns among village chicken populations across North, West and Central Africa. *BMC Genetics* **13**, 34.

Liu Y.P., Wu G.S., Yao Y.G., Miao Y.W., Luikart G., Baig M., Beja-Pereira A., Ding Z.L., Palanichamy M.G. and Zhang Y.P. (2006) Multiple maternal origins of chickens: out of the Asian jungles. *Molecular Phylogenetic Evolution* **38**, 12-19.

MAF (Ministry of Agriculture and Fisheries) (2013) Directorate of Rural Women; projects and programs of agricultural production and extension. Muscat, Oman.

Mtileni B.J., Muchadeyi F.C., Maiwashe A., Chimonyo M., Groeneveld E., Weigend S. and Dzama K. (2011a) Diversity and origin of South African chickens. *Poultry Sciences* **90**, 2189–2194.

Mtileni B.J., Muchadeyi F.C., Maiwashe A., Groeneveld E., Groeneveld L.F., Dzama K. and Weigend S. (2011b) Genetic diversity and conservation of South African indigenous chicken populations. *Journal of Animal Breeding and Genetics* **128**, 209-218.

Muchadeyi F.C., Eding H., Simianer H., Wollny C.B.A., Groeneveld E. and Weigend S. (2008) Mitochondrial DNA D-loop sequences suggest a Southeast Asian and Indian origin of Zimbabwean village chickens. *Animal Genetics* **39**, 615–622.

Muchadeyi F.C., Eding H., Wollny C.B., Groeneveld E., Makuza S.M., Shamseldin R., Simianer H. and Weigend S. (2007) Absence of population substructuring in Zimbabwe chicken ecotypes inferred using microsatellite analysis. *Animal Genetics* **38**, 332-339.

Mwacharo J.M., Bjornstad G., Mobegi V., Nomura K., Hanada H., Amano T., Jianlin H. and Hanotte O. (2011) Mitochondrial DNA reveals multiple introductions of domestic chicken in East Africa. *Molecular Phylogenetic Evolution* **58**, 374-382.

Mwacharo J.M., Nomura K., Hanada H., Han J.L., Amano T. and Hanotte O. (2013) Reconstructing the origin and dispersal patterns of village chickens across East Africa: insights from autosomal markers. *Molecular Ecology* **22**, 2683–2697.

Oliveira E.J., Pádua J.G., Zucchi M.I., Vencovsky R. and Vieira M.L.C. (2006) Origin , evolution and genome distribution of microsatellites. *Genetics and Molecular Biology* **29** (1), 294-307.

Pica-Ciamarra U. and Dhawan M. (2010) Small-scale poultry farming and poverty reduction in South Asia: from good practices to good policies in Bangladesh, Bhutan and India. In: *South-Asia Pro-Poor Livestock Policy Programme*. NDDB and FAO (Eds).

Potts D.T. (2000) *Ancient Magan, the secrets of Tell Abraq*. Trident Press Ltd., UK.

Rege J.E.O. (1999) The state of African cattle genetic resources I. Classification framework and identification of threatened and extinct breeds. *Animal Genetic Resources Information* **25**, 1-25.

Saleh F. (2000) Improving the local Dhofari poultry, Progress Report. Ministry of Agriculture and Fisheries, Dhofar, Sultanate of Oman.

Sarkar K. and Golam M. (2009) Commercial family poultry farming with local chickens: effective strategies for family poultry in Bangladesh. *World's Poultry Science Journal* **65**, 251-259.